T0214062

Developments in Mathematics

Volume 65

Christian Constanda • Dale Doty

The Generalized Fourier Series Method

Bending of Elastic Plates

 Springer

Christian Constanda
The Charles W. Oliphant
Professor of Mathematics
Department of Mathematics
The University of Tulsa
Tulsa, OK, USA

Dale Doty
Department of Mathematics
The University of Tulsa
Tulsa, OK, USA

ISSN 1389-2177   ISSN 2197-795X (electronic)
Developments in Mathematics
ISBN 978-3-030-55851-2   ISBN 978-3-030-55849-9 (eBook)
https://doi.org/10.1007/978-3-030-55849-9

Mathematics Subject Classification: 31A10, 35C15, 35J57, 74K20, 74G15

This Springer imprint is published by the registered company Springer Nature Switzerland AG
The registered company address is: Gewerbestrasse 11, 6330 Cham, Switzerland

For Lia
   and the newest generation:
   Jack, Emily, and Archie
            (CC)

For Jennifer
            (DD)

Preface

The process of bending of elastic plates is of considerable importance in a variety of applications, from civil engineering projects to ventures in the aerospace industry. Since the inception of its study, many mathematical models have been constructed and used for the qualitative and quantitative investigation of its salient features. Leaving aside the particulars of individual cases, these models can be classified into two main categories: Kirchhoff type [24] and Mindlin–Reissner type [27, 28]. The former, which reduces to the solution of one fourth-order partial differential equation, is easier to handle but produces only basic information. The latter consists generally of a system of three second-order equations and accompanying conditions, and is regarded as more refined in the sense that it trades the simplicity of the Kirchhoff model for additional details on the evolution of the physical phenomenon. However, the enhanced sophistication of the Mindlin–Reissner models is achieved at the expense of a certain amount of lack of mathematical consistency. By contrast, the model that forms the object of our attention, described in [15] and other published material, has no such drawback, being based exclusively on the averaging (across the thickness of the plate) of the exact equations of three-dimensional linear elasticity, coupled with an assumption on the form of the displacement vector.

Our model has been studied extensively in the literature. The equilibrium problem is treated in [15] and [1] (see also [14]), where the questions of existence, uniqueness, and integral representation of the solution are answered in full; the comprehensive discussion of the time-dependent case can be found in [2]; the harmonic plate oscillations are analyzed in [29]; thermoelastic issues are examined in [3–13]; and a boundary element method is developed in [23] for the deformation of a plate on an elastic foundation.

As all scientists and engineers know only too well, it is very seldom that the solution of a mathematical model can be expressed in closed form—that is, by means of a specific, well-defined set of functions. In general, we have to be satisfied instead with a numerical approximation, accurate enough to be acceptable for practical purposes. Our book develops precisely this type of procedure, which makes use of generalized Fourier series and can be applied to any mathematical model governed by an elliptic equation or system with constant coefficients. We chose this particular one not only because of its utilitarian value but also because of its complexity, which showcases the power, sophistication, and adaptability of the proposed technique.

The idea behind such a method was put forward earlier in [26], in the context of three-dimensional elasticity. However, that construction does not work for our model, which is governed by a "hybrid" system of three equations with three unknown functions in only two independent variables, and with a very complicated far-field pattern that requires a novel way of handling.

In Chap. 1, we gather together (without proof) the main results concerning the mathematical model, as developed in [15], and list the asymptotic expansions of various mathematical objects encountered in later analysis. Chapters 3–8 present the method in application to the interior and exterior Dirichlet, Neumann, and Robin boundary value problems. This is done in great detail, with comments on the theoretical aspects and their numerical implementation, especially for the first problem (Chap. 3), which is designed as a template that contains all the calculations and explanatory graphs supporting the mathematical and computational arguments. Some of these details, deemed more ancillary than essential, are not included in the other cases since the interested reader can develop them easily by following the patterns established for the interior Dirichlet problem.

Additional comments, together with a comprehensive guide to the use of the *Mathematica*® software, the main tool for the numerical segments in the book, can be found in the Appendix. Since we wanted to help the reader by making independent, self-contained presentations of each problem, we saw it necessary to insert some identical portions of explicative text in all of them. But these are short passages that, we hope, will not detract from the fluency of the narrative.

Preliminary brief announcements of some of the results in the book have appeared in [16–22].

Before turning our attention to the mathematical discourse itself, we would like to thank Elizabeth Loew, Executive Editor for Mathematics at Springer, New York, for her expert advice and her flexibility and understanding over deadlines.

Finally, we wish to express our deep gratitude to our wives, without whose unflinching support and selflessness this book would not have seen the light of day.

Tulsa, OK, USA   Christian Constanda
Tulsa, OK, USA   Dale Doty
June 2020

Contents

Chapter 1
The Mathematical Model

1.1 Basic Equations

We introduce some symbols and conventions that are used throughout the book.

1.1 Notation.

(i) Generic points in $\mathbb{R}^2$ are denoted by $x = (x_1, x_2)$ or $y = (y_1, y_2)$, and the Euclidean norm in $\mathbb{R}^2$ is denoted by $|\cdot|$.

(ii) S^+ is a finite domain in $\mathbb{R}^2$ (containing the origin) with a simple, closed, C^2 boundary curve ∂S, and $S^- = \mathbb{R}^2 \setminus (S^+ \cup \partial S)$. We write $\bar{S}^+ = S^+ \cup \partial S$ and $\bar{S}^- = S^- \cup \partial S$.

(iii) If $Q = (q_{ij})$, $i = 1, \ldots, m$, $j = 1, \ldots, n$, is an $m \times n$ matrix, then $Q_{(i)}$, $i = 1, 2, \ldots, m$, are its rows, $Q^{(j)}$, $j = 1, 2, \ldots, n$, are its columns, and Q^{T} is its transpose. We represent a matrix Q in terms of its columns as

$$Q = (Q^{(1)} \ Q^{(2)} \ \cdots \ Q^{(n)}).$$

(iv) If $Q = (q_{ij})$, $i = 1, \ldots, m$, $j = 1, \ldots, n$, is an $m \times n$ matrix function, X is a space of scalar functions, and L is an operator on X, we write $Q \in X$ if all the entries $q_{ij} \in X$, and designate by LQ the $m \times n$ matrix with entries Lq_{ij}.

(v) The symbol I is used for both the identity matrix and the identity operator, as determined by context.

(vi) When a function g defined on $\bar{S}^+$ or $\bar{S}^-$ occurs in a formula with its restriction to ∂S, we sometimes (but not always) use the same symbol g for it instead of $g|_{\partial S}$, the meaning being obvious from the context.

(vii) $C^{0,\alpha}(\partial S)$ and $C^{1,\alpha}(\partial S)$, $\alpha \in (0, 1)$, are the spaces of Hölder continuous functions and Hölder continuously differentiable functions on ∂S, respectively.

(viii) Let g be a continuous function on $S^+ \cup S^-$. We denote by $g^+(x)$ $(g^-(x))$ the limit, if it exists, of g as the point $x \in \partial S$ is approached from within S^+ (S^-).

© Springer Nature Switzerland AG 2020

C. Constanda, D. Doty, *The Generalized Fourier Series Method*, Developments in Mathematics 65, https://doi.org/10.1007/978-3-030-55849-9_1

(ix) A linear operator is referred to as an operator with a zero (nonzero) null space if its null space consists (does not consist) of the zero element alone.

(x) Greek subscripts take the values 1, 2, and summation over repeated such subscripts is understood; for example, $v_\alpha w_\alpha = v_1 w_1 + v_2 w_2$.

1.2 Definition.

(i) If $f, g \in L^2(\partial S)$ are scalar functions defined on ∂S, then the inner product of f and g and the norm of f are, respectively,

$$\langle f, g \rangle = \int_{\partial S} f g \, ds, \quad \|f\| = (\langle f, f \rangle)^{1/2}.$$

(ii) If $v, w \in L^2(\partial S)$ are $n \times 1$ column vector functions, we write

$$\langle v, w \rangle = \int_{\partial S} v^\mathsf{T} w \, ds.$$

(iii) If H is a $n \times n$ matrix function and v is a $n \times 1$ column vector function in $L^2(\partial S)$, we write

$$\langle H, v \rangle = \int_{\partial S} H v \, ds = \int_{\partial S} \begin{pmatrix} H_{(1)} v \\ H_{(2)} v \\ \vdots \\ H_{(n)} v \end{pmatrix} ds$$

$$= \int_{\partial S} \begin{pmatrix} \left((H^\mathsf{T})^{(1)} \right)^\mathsf{T} v \\ \left((H^\mathsf{T})^{(2)} \right)^\mathsf{T} v \\ \vdots \\ \left((H^\mathsf{T})^{(n)} \right)^\mathsf{T} v \end{pmatrix} ds = \begin{pmatrix} \langle (H^\mathsf{T})^{(1)}, v \rangle \\ \langle (H^\mathsf{T})^{(2)}, v \rangle \\ \vdots \\ \langle (H^\mathsf{T})^{(n)}, v \rangle \end{pmatrix}. \tag{1.1}$$

(iv) If $H, K \in L^2(\partial S)$ are $n \times n$ matrix functions, we follow the pattern in (1.1) and write

$$\langle H, K \rangle = \int_{\partial S} H K \, ds = \int_{\partial S} \begin{pmatrix} H_{(1)} K^{(1)} & H_{(1)} K^{(2)} & \cdots & H_{(1)} K^{(n)} \\ H_{(2)} K^{(1)} & H_{(2)} K^{(2)} & \cdots & H_{(2)} K^{(n)} \\ \vdots & \vdots & \vdots & \vdots \\ H_{(n)} K^{(1)} & H_{(n)} K^{(2)} & \cdots & H_{(n)} K^{(n)} \end{pmatrix} ds$$

$$= \begin{pmatrix} \langle (H^\mathsf{T})^{(1)}, K^{(1)} \rangle & \langle (H^\mathsf{T})^{(1)}, K^{(2)} \rangle & \cdots & \langle (H^\mathsf{T})^{(1)}, K^{(n)} \rangle \\ \langle (H^\mathsf{T})^{(2)}, K^{(1)} \rangle & \langle (H^\mathsf{T})^{(2)}, K^{(2)} \rangle & \cdots & \langle (H^\mathsf{T})^{(2)}, K^{(n)} \rangle \\ \vdots & \vdots & \vdots \\ \langle (H^\mathsf{T})^{(n)}, K^{(1)} \rangle & \langle (H^\mathsf{T})^{(n)}, K^{(2)} \rangle & \cdots & \langle (H^\mathsf{T})^{(n)}, K^{(n)} \rangle \end{pmatrix}$$

$$= \left(\langle (H^\mathsf{T})^{(i)}, K^{(j)} \rangle \right), \quad i, j = 1, 2, \ldots, n.$$

1.3 Remark. We recall two useful properties of matrix multiplication. If H and K are $n \times n$ matrices, then

$$(HK)_{(j)} = H_{(j)}K, \quad (HK)^{(j)} = HK^{(j)}. \tag{1.2}$$

A *thin plate* is an elastic body that occupies a region $\bar{S} \times [-h_0/2, h_0/2]$ in $\mathbb{R}^3$, where S is a (finite or infinite) domain in $\mathbb{R}^2$ with a smooth boundary ∂S and $0 < h_0 = \text{const} \ll \text{diam}\, S$. A detailed study of this type of object, with complete comments and proofs of all the mathematical statements, can be found in [14] and [15]. Here, we summarize the main results and describe additional features that facilitate the construction of our numerical approximation schemes. For convenience, in what follows we generally use the same notation as in [14] and [15].

Other theories of bending of elastic plates are briefly compared and contrasted in [15, pp. 2–3], with the one investigated in this book.

The equilibrium equations of our model in the absence of body forces and moments are

$$A(\partial_x)u(x) = 0, \tag{1.3}$$

where the matrix partial differential operator

$$A(\partial_x) = A(\partial/\partial x_1, \partial/\partial x_2))$$

is defined by

$$A(\xi_1, \xi_2) = \begin{pmatrix} h^2\mu\Delta + h^2(\lambda+\mu)\xi_1^2 - \mu & h^2(\lambda+\mu)\xi_1\xi_2 & -\mu\xi_1 \\ h^2(\lambda+\mu)\xi_1\xi_2 & h^2\mu\Delta + h^2(\lambda+\mu)\xi_2^2 - \mu & -\mu\xi_2 \\ \mu\xi_1 & \mu\xi_2 & \mu\Delta \end{pmatrix},$$

$u = (u_1, u_2, u_3)^\mathsf{T}$ is a vector characterizing the displacement field, $\Delta = \xi_\alpha\xi_\alpha$, λ and μ are the Lamé constants of the material, and $h = h_0/\sqrt{12}$.

The boundary bending and twisting moments and transverse shear force are expressed as Tu, where the operator

$$T(\partial_x) = T(\partial/\partial_{x_1}, \partial/\partial_{x_2})$$

is defined by

$$T(\xi_1, \xi_2) = \begin{pmatrix} h^2(\lambda+2\mu)n_1\xi_1 + h^2\mu n_2\xi_2 & h^2\mu n_2\xi_1 + h^2\lambda n_1\xi_2 & 0 \\ h^2\lambda n_2\xi_1 + h^2\mu n_1\xi_2 & h^2\mu n_1\xi_1 + h^2(\lambda+2\mu)n_2\xi_2 & 0 \\ \mu n_1 & \mu n_2 & \mu n_\alpha\xi_\alpha \end{pmatrix}$$

and $n = (n_1, n_2)^\mathsf{T}$ is the unit outward normal to ∂S.

The internal energy density per unit area of the middle plane is

$$E(u, u) = \tfrac{1}{2}\left\{ h^2\left[E_0(u, u) + \mu(u_{1,2} + u_{2,1})^2\right] \right.$$
$$\left. + \mu\left[(u_1 + u_{3,1})^2 + (u_2 + u_{3,2})^2\right] \right\},$$

where

$$E_0(u,u) = (\lambda + 2\mu)u_{1,1}^2 + 2\lambda u_{1,1}u_{2,2} + (\lambda + 2\mu)u_{2,2}^2.$$

Throughout what follows, we assume that the Lamé constants satisfy the conditions

$$\lambda + \mu > 0, \quad \mu > 0.$$

The statements listed in the rest of this chapter can be found with their complete proofs at the locations indicated in [15].

1.4 Theorem. ([15], Theorem 3.2, p. 69) *$E(u,u)$ is a positive quadratic form and* (1.3) *is an elliptic system.*

1.5 Theorem. ([15], Theorem 3.3, p. 70) *$E(u,u) = 0$ if and only if*

$$u(x) = (c_1, c_2, c_0 - c_1 x_1 - c_2 x_2)^\mathsf{T}, \tag{1.4}$$

where $c_0, c_\alpha = $ const.

It is obvious that the columns of the matrix

$$F = \begin{pmatrix} 1 & 0 & 0 \\ 0 & 1 & 0 \\ -x_1 & -x_2 & 1 \end{pmatrix} \tag{1.5}$$

form a basis for the vector space $\mathscr{F}$ of rigid displacements, and that a generic rigid displacement of the form (1.4) can be written as

$$u = Fc, \tag{1.6}$$

where

$$c = (c_1, c_2, c_0)^\mathsf{T}.$$

1.6 Remarks.

(i) To avoid notation clashes later on, in some parts of the book we use the symbol Fa instead of Fc.

(ii) The space of the restrictions of the elements of $\mathscr{F}$ to ∂S is denoted by $\mathscr{F}|_{\partial S}$.

1.7 Theorem. ([15], Theorem 3.6, p. 72) *If $u \in C^2(S^+) \cap C^1(\bar{S}^+)$ is a solution of system* (1.3) *in S^+, then*

$$2 \int_{S^+} E(u,u)\,da = \int_{\partial S} u^\mathsf{T} Tu\,ds.$$

We construct the matrix of fundamental solutions ([2], p. 73)

$$D(x,y) = A^*(\partial_x)t(x,y) \tag{1.7}$$

for the operator $-A$, where $A^*(\xi_1, \xi_2)$ is the adjoint of the matrix $A(\xi_1, \xi_2)$,

$$t(x,y) = t(|x-y|) = a_2\big[(4h^2+|x-y|^2)\ln|x-y| + 4h^2K_0(h^{-1}|x-y|)\big], \quad (1.8)$$

K_0 is the modified Bessel function of order zero, and

$$a_2 = \big(8\pi h^2\mu^2(\lambda+2\mu)\big)^{-1}.$$

It can be verified that

$$D(x,y) = \big(D(y,x)\big)^{\mathsf{T}}. \tag{1.9}$$

We also introduce the matrix of singular solutions ([15], formula (3.28), p. 74)

$$P(x,y) = \big(T(\partial_y)D(y,x)\big)^{\mathsf{T}}. \tag{1.10}$$

1.8 Remark. When the boundary operator T is applied to a vector (matrix) function of only one point-variable, that variable is not mentioned explicitly in the notation for T. For example, if $\bar{x}$ is a specified (fixed) point, then we write

$$T(\partial_x)D(x,\bar{x}) = TD(x,\bar{x}).$$

From (1.7) and (1.8) we find ([15], formula (3.30), p. 75)) that for $y \in \partial S$ and x close to y,

$$D_{11}(x,y),\, D_{22}(x,y),\, D_{33}(x,y) = O(\ln|x-y|). \tag{1.11}$$

1.9 Theorem. ([2], Theorem 3.8, p. 76) *The columns of $D(x,y)$ and $P(x,y)$ are solutions of system* (1.3) *at all* $x \in \mathbb{R}^2$, $x \neq y$.

1.10 Theorem. ([15], Theorem 3.9, p. 76) *If* $u \in C^2(S^+)\cap C^1(\bar{S}^+)$ *is a solution of system* (1.3) *in* S^+, *then*

$$\int_{\partial S}\big[D(x,y)Tu(y) - P(x,y)u(y)\big]\,ds(y) = \begin{cases} u(x), & x \in S^+, \\ \tfrac{1}{2}u(x), & x \in \partial S, \\ 0, & x \in S^-. \end{cases} \tag{1.12}$$

1.11 Remark. It is not difficult to extract the rigid displacement from a known solution of system (1.3) in S^+. We write the linear part of the solution in the form

$$v = f + w, \tag{1.13}$$

where f is the rigid displacement component and w is the linear residual that contains no rigid displacement. Then

$$Av = Af + Aw = Aw,$$

from which we determine w, generally, as it turns out, in terms of three arbitrary constants. Suppressing the rigid displacement introduced by these constants in the expression of w, we now use (1.13) to obtain

$$f = v - w.$$

1.2 Far Field Behavior

For y fixed and $|x| \to \infty$, we have the following basic asymptotic expansions, constructed by direct calculation:

$$\frac{1}{|x-y|^2} = \frac{1}{|x|^2} + 2\frac{\langle x,y \rangle}{|x|^4} - \frac{|y|^2}{|x|^4} + 4\frac{\langle x,y \rangle^2}{|x|^6} + O(|x|^{-5}),$$

$$\ln|x-y| = \ln|x| - \frac{\langle x,y \rangle}{|x|^2} + \frac{1}{2}\frac{|y|^2}{|x|^2} - \frac{\langle x,y \rangle^2}{|x|^4}$$
$$+ \frac{\langle x,y \rangle|y|^2}{|x|^4} - \frac{4}{3}\frac{\langle x,y \rangle^3}{|x|^6} + O(|x|^{-4}), \qquad (1.14)$$

$$K_0\left(h^{-1}|x-y|\right) = O\left(|x|^{-1/2}e^{-|x|}\right).$$

Let $\mathscr{A}$ be the set of 3×1 vector functions u in S^- that, in terms of polar coordinates r, θ with the pole at the origin, admit, as $r \to \infty$, an asymptotic expansion of the form (see [15], formula (3.39), p. 78)

$$u_1(r,\theta) = r^{-1}\left[m_0\sin\theta + 2m_1\cos\theta - m_0\sin(3\theta) + (m_2 - m_1)\cos(3\theta)\right]$$
$$+ r^{-2}\left[(2m_3 + m_4)\sin(2\theta) + m_5\cos(2\theta) - 2m_3\sin(4\theta) + 2m_6\cos(4\theta)\right]$$
$$+ r^{-3}\left[2m_7\sin(3\theta) + 2m_8\cos(3\theta) + 3(m_9 - m_7)\sin(5\theta)\right.$$
$$\left. + 3(m_{10} - m_8)\cos(5\theta)\right] + O(r^{-4}),$$

$$u_2(r,\theta) = r^{-1}\left[2m_2\sin\theta + m_0\cos\theta + (m_2 - m_1)\sin(3\theta) + m_0\cos(3\theta)\right]$$
$$+ r^{-2}\left[(2m_6 + m_5)\sin(2\theta) - m_4\cos(2\theta) + 2m_6\sin(4\theta) + 2m_3\cos(4\theta)\right]$$
$$+ r^{-3}\left[2m_{10}\sin(3\theta) - 2m_9\cos(3\theta) + 3(m_{10} - m_8)\sin(5\theta)\right. \qquad (1.15)$$
$$\left. + 3(m_7 - m_9)\cos(5\theta)\right] + O(r^{-4}),$$

$$u_3(r,\theta) = -(m_1 + m_2)\ln r - \left[m_1 + m_2 + m_0\sin(2\theta) + (m_1 - m_2)\cos(2\theta)\right]$$
$$+ r^{-1}\left[(m_3 + m_4)\sin\theta + (m_5 + m_6)\cos\theta - m_3\sin(3\theta) + m_6\cos(3\theta)\right]$$
$$+ r^{-2}\left[m_{11}\sin(2\theta) + m_{12}\cos(2\theta) + (m_9 - m_7)\sin(4\theta)\right.$$
$$\left. + (m_{10} - m_8)\cos(4\theta)\right] + O(r^{-3}),$$

where $m_1, \ldots, m_{12}$ are arbitrary constants.

We also introduce the set

$$\mathscr{A}^* = \left\{u^* \mid u^* = u + u_0, \; u \in \mathscr{A}, \; u_0 \text{ is of the form (1.6)}\right\}$$

and adopt the notation $x_p = (r, \theta)$.

1.12 Remark. We note that the term $-(m_1 + m_2)$ in the third entry in (1.15) represents a vertical rigid displacement (translation).

1.13 Remark. Since all asymptotic expansions in what follows are performed exclusively on two-point 3×1 vector functions in S^- of the form $v(x, y)$ for y fixed and $|x| \to \infty$, this detail will not be mentioned explicitly in any of our calculations.

Obviously, not all such functions are of class $\mathscr{A}$. Some, however, when expanded asymptotically as explained above, can be written as the sum of a class $\mathscr{A}$ component and a residual that consists of finitely many terms. For such functions v, we write

$$v(x, y) = \left(v(x, y) \right)^{\mathscr{A}} + \left(v(x, y) \right)^{\infty},$$

where the notation is self-explanatory. Since the expansion of v is carried out entry by entry, we have

$$\left((v(x, y)) \right)_i = \left((v(x, y))^{\mathscr{A}} \right)_i + \left((v(x, y))^{\infty} \right)_i,$$

with

$$
\begin{aligned}
\left((v(x, y))^{\mathscr{A}} \right)_i &= \left((v(x, y))_i \right)^{\mathscr{A}}, \\
\left((v(x, y))^{\infty} \right)_i &= \left((v(x, y))_i \right)^{\infty};
\end{aligned}
\tag{1.16}
$$

here, $\left((v(x, y))_i \right)^{\mathscr{A}}$ has the form of the expansion for $u_i(r, \theta)$ in (1.15).

If the vector function v depends only on x, then we simplify the notation by writing

$$\left(v(x) \right)^{\mathscr{A}} = v^{\mathscr{A}}(x).$$

1.14 Definition. For the sake of brevity, we refer to the process of constructing the class $\mathscr{A}$ component of a function v by means of the asymptotic expansion described in Remark 1.13, as the $\mathscr{A}$-*expansion* of v. In the case of a matrix, this applies to each of its columns, with the expansion performed on the individual entries. In an $\mathscr{A}$-expansion, x is normally understood as, and frequently replaced by, x_p.

1.2.1 Asymptotic Expansion of Matrix $D(x, y)$

1.15 Remark. As we see in this subsection, direct computation by means of formulas (1.7), (1.8), and (1.14) of the far field for $D(x, y)$ shows that each column $(D(x, y))^{(j)}$ of this matrix function is the sum of a class $\mathscr{A}$ component $\left((D(x, y))^{(j)} \right)^{\mathscr{A}}$ and a residual $\left((D(x, y))^{(j)} \right)^{\infty}$, with the latter consisting of finitely many terms. Thus, we write

$$D(x, y) = \left(D(x, y) \right)^{\mathscr{A}} + \left(D(x, y) \right)^{\infty},$$

where the two terms on the right-hand side are the matrix functions with columns $\left((D(x, y))^{(j)} \right)^{\mathscr{A}}$ and $\left((D(x, y))^{(j)} \right)^{\infty}$, respectively; that is,

$$\left(\left(D(x,y) \right)^{\mathscr{A}} \right)^{(j)} = \left(\left(D(x,y) \right)^{(j)} \right)^{\mathscr{A}},$$
$$\left(\left(D(x,y) \right)^{\infty} \right)^{(j)} = \left(\left(D(x,y) \right)^{(j)} \right)^{\infty}.$$

In accordance with Remark 1.13, we also write

$$\left(D(x,y) \right)_{ij} = \left(\left(D(x,y) \right)_{ij} \right)^{\mathscr{A}} + \left(\left(D(x,y) \right)_{ij} \right)^{\infty},$$

where, by (1.16),

$$\left(\left(D(x,y) \right)_{ij} \right)^{\mathscr{A}} = \left(\left(\left(D(x,y) \right)^{(j)} \right)_i \right)^{\mathscr{A}} = \left(\left(\left(D(x,y) \right)^{(j)} \right)^{\mathscr{A}} \right)_i$$
$$= \left(\left(\left(D(x,y) \right)^{\mathscr{A}} \right)^{(j)} \right)_i = \left(\left(D(x,y) \right)^{\mathscr{A}} \right)_{ij}. \qquad (1.17)$$

Similarly,

$$\left(\left(D(x,y) \right)_{ij} \right)^{\infty} = \left(\left(D(x,y) \right)^{\infty} \right)_{ij}.$$

Our calculations involve a few other matrices constructed from $D(x,y)$, which include a superscript $\mathscr{A}$ somewhere in their notation. This means that their individual entries have an $\mathscr{A}$-expansion that coincides with one of the entries in (1.15), but does not automatically imply that their columns belong to class $\mathscr{A}$. The use of the superscript $\mathscr{A}$ in such cases is necessary to show how these matrices are designed.

The asymptotic expansion of the first entry of $\left(D(x,y) \right)^{(1)}$ in Cartesian coordinates is

$$
\begin{aligned}
&\left(D(x,y) \right)_{11} \\
&= a_2 \mu \Bigg\{ -\mu (2 \ln r + 1) - 2\mu \frac{x_1^2}{|x|^2} \\
&\quad + 2\mu \left(3 \frac{x_1}{|x|^2} y_1 + \frac{x_2}{|x|^2} y_2 - 2 \frac{x_1^3}{|x|^4} y_1 - 2 \frac{x_1^2 x_2}{|x|^4} y_2 \right) \\
&\quad - 4h^2 (\lambda + 2\mu) \left(\frac{1}{|x|^2} - 2 \frac{x_1^2}{|x|^4} \right) \\
&\quad - \mu \Bigg[\frac{1}{|x|^2} (3y_1^2 + y_2^2) - 2 \frac{x_1^2}{|x|^4} (6y_1^2 + y_2^2) - 12 \frac{x_1 x_2}{|x|^4} y_1 y_2 \\
&\qquad - 2 \frac{x_2^2}{|x|^4} y_2^2 + 8 \frac{x_1^4}{|x|^6} y_1^2 + 16 \frac{x_1^3 x_2}{|x|^6} y_1 y_2 + 8 \frac{x_1^2 x_2^2}{|x|^6} y_2^2 \Bigg] \\
&\quad + 8h^2 (\lambda + 2\mu) \left(-3 \frac{x_1}{|x|^4} y_1 - \frac{x_2}{|x|^4} y_2 + 4 \frac{x_1^3}{|x|^6} y_1 + 4 \frac{x_1^2 x_2}{|x|^6} y_2 \right) \\
&\quad - 2\mu \Bigg[\frac{x_1}{|x|^4} (5y_1^3 + 3y_1 y_2^2) + \frac{x_2}{|x|^4} (3y_1^2 y_2 + y_2^3) \\
&\qquad - 4 \frac{x_1^3}{|x|^6} \left(\frac{10}{3} y_1^3 + y_1 y_2^2 \right) - 4 \frac{x_1^2 x_2}{|x|^6} (6y_1^2 y_2 + y_2^3)
\end{aligned}
$$

$$-12\frac{x_1x_2^2}{|x|^6}y_1y_2^2-\frac{4}{3}\frac{x_2^3}{|x|^6}y_2^3+8\frac{x_1^5}{|x|^8}y_1^3$$

$$+24\frac{x_1^4x_2}{|x|^8}y_1^2y_2+24\frac{x_1^3x_2^2}{|x|^8}y_1y_2^2+8\frac{x_1^2x_2^3}{|x|^8}y_2^3\Big]\Big\}+O(|x|^4),$$

with class $\mathscr{A}$ component

$$\left((D(x_p,y))_{11}\right)^{\mathscr{A}}$$

$$=a_2\Big\{-\mu^2[2(\ln r+1)+\cos(2\theta)]$$

$$+r^{-1}[\mu^2y_2\sin\theta+3\mu^2y_1\cos\theta-\mu^2y_2\sin(3\theta)-\mu^2y_1\cos(3\theta)]$$

$$+r^{-2}[2\mu^2y_1y_2\sin(2\theta)+\left(4h^2\mu(\lambda+2\mu)+2\mu^2y_1^2\right)\cos(2\theta)$$

$$-2\mu^2y_1y_2\sin(4\theta)+\mu^2(y_2^2-y_1^2)\cos(4\theta)]$$

$$+r^{-3}\Big[\left(8h^2\mu(\lambda+2\mu)y_2+3\mu^2y_1^2y_2+\tfrac{1}{3}\mu^2y_2^3\right)\sin(3\theta)$$

$$+\left(8h^2\mu(\lambda+2\mu)y_1+\tfrac{5}{3}\mu^2y_1^3-\mu^2y_1y_2\right)\cos(3\theta)$$

$$+\mu^2(y_2^3-3y_1^2y_2)\sin(5\theta)+\mu^2(3y_1y_2^2-y_1^3)\cos(5\theta)\Big]\Big\}+O(r^{-4})$$

and residual

$$\left((D(x,y))_{11}\right)^{\infty}=-a_2\mu^2\left\{\frac{x_1^2-x_2^2}{x_1^2+x_2^2}+2+\ln(x_1^2+x_2^2)\right\},$$

or, in polar coordinates,

$$\left((D(x_p,y))_{11}\right)^{\infty}=-a_2\mu^2[2(\ln r+1)+\cos(2\theta)].$$

The asymptotic expansion of the second entry of $D^{(1)}$ in Cartesian coordinates is

$$(D(x,y))_{21}$$

$$=a_2\mu\Big\{-2\mu\frac{x_1x_2}{|x|^2}$$

$$+2\mu\left(\frac{x_1}{|x|^2}y_2+\frac{x_2}{|x|^2}y_1-2\frac{x_1^2x_2}{|x|^4}y_1-2\frac{x_1x_2^2}{|x|^4}y_2\right)$$

$$+8h^2(\lambda+2\mu)\frac{x_1x_2}{|x|^4}$$

$$+2\mu\Big[\frac{1}{|x|^2}y_1y_2+3\frac{x_1x_2}{|x|^4}(y_1^2+y_2^2)$$

$$-4\frac{x_1^3x_2}{|x|^6}y_1^2-8\frac{x_1^2x_2^2}{|x|^6}y_1y_2-4\frac{x_1x_2^3}{|x|^6}y_2^2\Big]$$

$$
+8h^2(\lambda+2\mu)\left(-\frac{x_1}{|x|^4}y_2-\frac{x_2}{|x|^4}y_1+4\frac{x_1^2x_2}{|x|^6}y_1+4\frac{x_1x_2^2}{|x|^6}y_2\right)
$$

$$
-2\mu\left[\frac{x_1}{|x|^4}(3y_1^2y_2+y_2^3)+\frac{x_2}{|x|^4}(3y_1y_2^2+y_1^3)-4\frac{x_1^3}{|x|^6}y_1^2y_2\right.
$$

$$
-4\frac{x_1^2x_2}{|x|^6}(2y_1^3+3y_1y_2^2)-4\frac{x_1x_2^2}{|x|^6}(3y_1^2y_2+2y_2^3)-4\frac{x_2^3}{|x|^6}y_1y_2^2
$$

$$
\left.+8\frac{x_1^4x_2}{|x|^8}y_1^3+24\frac{x_1^3x_2^2}{|x|^8}y_1^2y_2+24\frac{x_1^2x_2^3}{|x|^8}y_1y_2^2+8\frac{x_1x_2^4}{|x|^8}y_2^3\right]\right\}+O(|x|^{-4}),
$$

with class $\mathscr{A}$ component

$$
\left((D(x_p,y))_{21}\right)^{\mathscr{A}}
$$
$$
=a_2\{-\mu^2\sin(2\theta)
$$
$$
+r^{-1}[\mu^2y_1\sin\theta+\mu^2y_2\cos\theta-\mu^2y_1\sin(3\theta)+\mu^2y_2\cos(3\theta)]
$$
$$
+r^{-2}[(4h^2\mu(\lambda+2\mu)+\mu^2(y_1^2+y_2^2))\sin(2\theta)
$$
$$
+\mu^2(y_2^2-y_1^2)\sin(4\theta)+2\mu^2y_1y_2\cos(4\theta)]
$$
$$
+r^{-3}[(8h^2\mu(\lambda+2\mu)y_1+\mu^2(y_1^3+y_1y_2^2))\sin(3\theta)
$$
$$
-(8h^2\mu(\lambda+2\mu)y_2+\mu^2(y_1^2y_2+y_2^3))\cos(3\theta)
$$
$$
+\mu^2(3y_1y_2^2-y_1^3)\sin(5\theta)+\mu^2(3y_1^2y_2-y_2^3)\cos(5\theta)]\}+O(r^{-4})
$$

and residual
$$
\left((D(x,y))_{21}\right)^{\infty}=-2a_2\mu^2\frac{x_1x_2}{x_1^2+x_2^2},
$$

or, in polar coordinates,

$$
\left((D(x_p,y))_{21}\right)^{\infty}=-a_2\mu^2\sin(2\theta).
$$

The asymptotic expansion of the third entry of $D^{(1)}$ in Cartesian coordinates is

$$
(D(x,y))_{31}
$$
$$
=-a_2\left\{\mu^2(2\ln|x|+1)y_1-\mu^2((2\ln|x|+1)x_1\right.
$$
$$
+2\mu^2\left(\frac{x_1}{|x|^2}y_1+\frac{x_1x_2}{|x|^2}y_2\right)
$$
$$
+2\mu^2\left[-\frac{x_1}{|x|^2}\left(\frac{3}{2}y_1^2+\frac{1}{2}y_2^2\right)-\frac{x_2}{|x|^2}y_1y_2\right.
$$
$$
\left.+\frac{x_1^3}{|x|^4}y_1^2+2\frac{x_1^2x_2}{|x|^4}y_1y_2+\frac{x_1x_2^2}{|x|^4}y_2^2\right]
$$
$$
+\mu^2\left[\frac{1}{|x|^2}(y_1^3+y_1y_2^2)-2\frac{x_1^2}{|x|^4}(2y_1^3+y_1y_2^2)\right.
$$

$$- 2\frac{x_1 x_2}{|x|^4}(3y_1^2 y_2 + y_2^3) - 2\frac{x_2^2}{|x|^4} y_1 y_2^2 + \frac{8}{3}\frac{x_1^4}{|x|^6} y_1^3$$

$$+ 8\frac{x_1^3 x_2}{|x|^6} y_1^2 y_2 + 8\frac{x_1^2 x_2^2}{|x|^6} y_1 y_2^2 + \frac{8}{3}\frac{x_1 x_2^3}{|x|^6} y_2^3 \Big] \Big\} + O(|x|^{-3}),$$

with class $\mathscr{A}$ component

$$\big((D(x_p, y))_{31}\big)^{\mathscr{A}}$$
$$= a_2 \Big\{ [\mu^2 r(2\ln r + 1) - 4h^2 \mu(\lambda + 2\mu)r^{-1}]\cos\theta$$
$$\quad - 2\mu^2 y_1 \ln r - [2\mu^2 y_1 + \mu^2 y_2 \sin(2\theta) + \mu^2 y_1 \cos(2\theta)]$$
$$\quad + r^{-1}\Big[\mu^2 y_1 y_2 \sin\theta + \big(\tfrac{1}{2}\mu^2(3y_1^2 + y_2^2) + 4h^2 \mu(\lambda + 2\mu)\big)\cos\theta$$
$$\quad - \mu^2 y_1 y_2 \sin(3\theta) + \tfrac{1}{2}\mu^2(y_2^2 - y_1^2)\cos(3\theta)\Big]$$
$$\quad + r^{-2}\Big[\mu^2\big(y_1^2 y_2 + \tfrac{1}{3}y_2^3\big)\sin(2\theta) + \tfrac{2}{3}\mu^2 y_1^3 \cos(2\theta)$$
$$\quad + \mu^2\big(\tfrac{1}{3}y_2^3 - y_1^2 y_2\big)\sin(4\theta) + \mu^2\big(y_1 y_2^2 - \tfrac{1}{3}y_1^3\big)\cos(4\theta)\Big]\Big\} + O(r^{-3})$$

and residual

$$\big((D(x, y))_{31}\big)^{\infty} = a_2\Big\{ -4h^2 \mu(\lambda + 2\mu)\frac{x_1}{x_1^2 + x_2^2} + \mu^2 x_1 [1 + \ln(x_1^2 + x_2^2)]\Big\},$$

or, in polar coordinates,

$$\big((D(x_p, y))_{31}\big)^{\infty} = a_2[\mu^2 r(2\ln r + 1) - 4h^2 \mu(\lambda + 2\mu)r^{-1}]\cos\theta.$$

1.16 Remark. The above expansions show that the coefficients $m_0, \ldots, m_{12}$ of the far-field pattern (1.15) for $(D^{(1)})^{\mathscr{A}}(x_p, y)$ are

$$m_0 = a_2\mu^2 y_2,$$
$$m_1 = \tfrac{3}{2}a_2\mu^2 y_1,$$
$$m_2 = \tfrac{1}{2}a_2\mu^2 y_1,$$
$$m_3 = a_2\mu^2 y_1 y_2,$$
$$m_4 = 0,$$
$$m_5 = a_2[4h^2\mu(\lambda + 2\mu) + 2\mu^2 y_1^2],$$
$$m_6 = \tfrac{1}{2}a_2\mu^2(y_2^2 - y_1^2),$$
$$m_7 = a_2\Big[4h^2\mu(\lambda + 2\mu)y_2 + \tfrac{3}{2}\mu^2 y_1^2 y_2 + \tfrac{1}{6}\mu^2 y_2^3\Big],$$
$$m_8 = a_2\Big[4h^2\mu(\lambda + 2\mu)y_1 + \tfrac{5}{6}\mu^2 y_1^3 - \tfrac{1}{2}\mu^2 y_1 y_2^2\Big],$$
$$m_9 = a_2\Big[4h^2\mu(\lambda + 2\mu)y_2 + \tfrac{1}{2}\mu^2(y_1^2 y_2 + y_2^3)\Big],$$

$$m_{10} = a_2 \left[4h^2 \mu(\lambda + 2\mu) y_1 + \tfrac{1}{2} \mu^2 (y_1^3 + y_1 y_2^2) \right],$$
$$m_{11} = \mu^2 \left(y_1^2 y_2 + \tfrac{1}{3} y_2^3 \right),$$
$$m_{12} = \tfrac{2}{3} \mu^2 y_1^3.$$

The asymptotic expansion of the first entry of $D^{(2)}$ in Cartesian coordinates is

$$
\begin{aligned}
\left(D(x,y) \right)_{12} &= a_2 \mu \Bigg\{ -2\mu \frac{x_1 x_2}{|x|^2} \\
&+ 2\mu \left(\frac{x_1}{|x|^2} y_2 + \frac{x_2}{|x|^2} y_1 - 2\frac{x_1^2 x_2}{|x|^4} y_1 - 2\frac{x_1 x_2^2}{|x|^4} y_2 \right) \\
&+ 8h^2(\lambda + 2\mu) \frac{x_1 x_2}{|x|^4} \\
&- 2\mu \left[-\frac{1}{|x|^2} y_1 y_2 - 3\frac{x_1 x_2}{|x|^4}(y_1^2 + y_2^2) \right. \\
&\quad \left. + 4\frac{x_1^3 x_2}{|x|^6} y_1^2 + 8\frac{x_1^2 x_2^2}{|x|^6} y_1 y_2 + 4\frac{x_1 x_2^3}{|x|^6} y_2^2 \right] \\
&+ 8h^2(\lambda + 2\mu) \left(-\frac{x_1}{|x|^4} y_2 - \frac{x_2}{|x|^4} y_1 + 4\frac{x_1^2 x_2}{|x|^6} y_1 + 4\frac{x_1 x_2^2}{|x|^6} y_2 \right) \\
&- 2\mu \left[\frac{x_1}{|x|^4}(3y_1^2 y_2 + y_2^3) + \frac{x_2}{|x|^4}(3y_1 y_2^2 + y_1^3) - 4\frac{x_1^3}{|x|^6} y_1^2 y_2 \right. \\
&\quad - 4\frac{x_1^2 x_2}{|x|^6}(2y_1^3 + 3y_1 y_2^2) - 4\frac{x_1 x_2^2}{|x|^6}(3y_1^2 y_2 + 2y_2^3) \\
&\quad - 4\frac{x_2^3}{|x|^6} y_1 y_2^2 + 8\frac{x_1^4 x_2}{|x|^8} y_1^3 + 24\frac{x_1^3 x_2^2}{|x|^8} y_1^2 y_2 \\
&\quad \left. + 24\frac{x_1^2 x_2^3}{|x|^8} y_1 y_2^2 + 8\frac{x_1 x_2^4}{|x|^8} y_2^3 \right] \Bigg\} + O(|x|^{-4}),
\end{aligned}
$$

with class $\mathscr{A}$ component

$$
\begin{aligned}
\left((D(x_p, y))_{12} \right)^{\mathscr{A}} &= a_2 \{ -\mu^2 \sin(2\theta) \\
&+ r^{-1} [\mu^2 y_1 \sin\theta + \mu^2 y_2 \cos\theta - \mu^2 y_1 \sin(3\theta) + \mu^2 y_2 \cos(3\theta)] \\
&+ r^{-2} [((4h^2 \mu(\lambda + 2\mu) + \mu^2(y_1^2 + y_2^2)) \sin(2\theta) \\
&\quad + \mu^2(y_2^2 - y_1^2) \sin(4\theta) + 2\mu^2 y_1 y_2 \cos(4\theta)] \\
&+ r^{-3} [(8h^2 \mu(\lambda + 2\mu) y_1 + \mu^2(y_1^3 + y_1 y_2^2)) \sin(3\theta)
\end{aligned}
$$

$$- (8h^2\mu(\lambda + 2\mu)y_2 + \mu^2(y_1^2 y_2 + y_2^3))\cos(3\theta)$$
$$+ \mu^2(3y_1 y_2^2 - y_1^3)\sin(5\theta) + \mu^2(3y_1^2 y_2 - y_2^3)\cos(5\theta)]\} + O(r^{-4})$$

and residual

$$\big((D(x,y))_{12}\big)^\infty = -2a_2\mu^2\,\frac{x_1 x_2}{x_1^2 + x_2^2},$$

or, in polar coordinates,

$$\big((D(x_p,y))_{12}\big)^\infty = -a_2\mu^2 \sin(2\theta).$$

The asymptotic expansion of the second entry of $D^{(2)}$ in Cartesian coordinates is

$$(D(x,y))_{22}$$
$$= a_2\mu\Bigg\{ -\mu(2\ln|x| + 1) - 2\mu\frac{x_2^2}{|x|^2}$$
$$+ 2\mu\left(\frac{x_1}{|x|^2|}y_1 + 3\frac{x_2}{|x|^2}y_2 - 2\frac{x_1 x_2^2}{|x|^4}y_1 - 2\frac{x_2^3}{|x|^4}y_2\right)$$
$$- 4h^2(\lambda + 2\mu)\left(\frac{1}{|x|^2} - 2\frac{x_2^2}{|x|^4}\right)$$
$$- \mu\Bigg[\frac{1}{|x|^2}(y_1^2 + 3y_2^2) - 2\frac{x_1^2}{|x|^4}y_1^2 - 12\frac{x_1 x_2}{|x|^4}y_1 y_2$$
$$- 2\frac{x_2^2}{|x|^4}(y_1^2 + 6y_2^2) + 8\frac{x_1^2 x_2^2}{|x|^6}y_1^2 + 16\frac{x_1 x_2^3}{|x|^6}y_1 y_2 + 8\frac{x_2^4}{|x|^6}y_2^2\Bigg]$$
$$+ 8h^2(\lambda + 2\mu)\left(-\frac{x_1}{|x|^4}y_1 - 3\frac{x_2}{|x|^4}y_2 + 4\frac{x_1 x_2^2}{|x|^6}y_1 + 4\frac{x_2^3}{|x|^6}y_2\right)$$
$$- 2\mu\Bigg[\frac{x_1}{|x|^4}(y_1^3 + 3y_1 y_2^2) + \frac{x_2}{|x|^4}(3y_1^2 y_2 + 5y_2^3) - \frac{4}{3}\frac{x_1^3}{|x|^6}y_1^3$$
$$- 12\frac{x_1^2 x_2}{|x|^6}y_1^2 y_2 - 4\frac{x_1 x_2^2}{|x|^6}(y_1^3 + 6y_1 y_2^2)$$
$$- \frac{4}{3}\frac{x_2^3}{|x|^6}(3y_1^2 y_2 + 10y_2^3) + 8\frac{x_1^3 x_2}{|x|^8}y_1^3 + 24\frac{x_1^2 x_2^2}{|x|^8}y_1^2 y_2$$
$$+ 24\frac{x_1 x_2^4}{|x|^8}y_1 y_2^2 + 8\frac{x_2^5}{|x|^8}y_2^3\Bigg]\Bigg\} + O(|x|^{-4}),$$

with class $\mathscr{A}$ component

$$\big((D(x_p,y))_{22}\big)^{\mathscr{A}}$$
$$= a_2\{-\mu^2[2(\ln r + 1) - \cos(2\theta)]$$
$$+ r^{-1}[3\mu^2 y_2 \sin\theta + \mu^2 y_1 \cos\theta + \mu^2 y_2 \sin(3\theta) + \mu^2 y_1 \cos(3\theta)]$$

$$+r^{-2}[2\mu^2 y_1 y_2 \sin(2\theta) - (4h^2\mu(\lambda+2\mu)+2\mu^2 y_2^2)\cos(2\theta)$$
$$+2\mu^2 y_1 y_2 \sin(4\theta) + \mu^2(y_1^2 - y_2^2)\cos(4\theta)]$$
$$+r^{-3}\left[\left(-8h^2\mu(\lambda+2\mu)y_2 + \mu^2 y_1^2 y_2 - \tfrac{5}{3}\mu^2 y_2^3\right)\sin(3\theta)\right.$$
$$+\left(8h^2\mu(\lambda+2\mu)y_1 + \tfrac{1}{3}\mu^2 y_1^3 + 3\mu^2 y_1 y_2^2\right)\cos(3\theta)$$
$$\left.+\mu^2(3y_1^2 y_2 - y_2^3)\sin(5\theta) + \mu^2(y_1^3 - 3y_1 y_2^2)\cos(5\theta)\right]\} + O(r^{-4})$$

and residual

$$\left((D(x,y))_{22}\right)^\infty = -a_2\mu^2\left\{\frac{-x_1^2 + x_2^2}{x_1^2 + x_2^2} + 2 + \ln(x_1^2 + x_2^2)\right\},$$

or, in polar coordinates,

$$\left((D(x_p,y))_{22}\right)^\infty = -a_2\mu^2[2(\ln r + 1) - \cos(2\theta)].$$

The asymptotic expansion of the third entry of $D^{(2)}$ in Cartesian coordinates is

$$(D(x,y))_{32}$$
$$= a_2\left\{\mu^2(2\ln|x|+1)(x_2 - y_2)\right.$$
$$-2\mu^2\left(\frac{x_1 x_2}{|x|^2}y_1 + \frac{x_2^2}{|x|^2}y_2\right)$$
$$+2\mu^2\left[\frac{x_1}{|x|^2}y_1 y_2 + \frac{1}{2}\frac{x_2}{|x|^2}(y_1^2 + 3y_2^2)\right.$$
$$\left.-\frac{x_1^2 x_2}{|x|^4}y_1^2 - 2\frac{x_1 x_2^2}{|x|^4}y_1 y_2 - \frac{x_2^3}{|x|^4}y_2^2\right]$$
$$+\mu^2\left[-\frac{1}{|x|^2}(y_1^2 y_2 + y_2^3) + 2\frac{x_1^2}{|x|^4}y_1^2 y_2\right.$$
$$+2\frac{x_1 x_2}{|x|^4}(y_1^3 + +3y_1 y_2^2) + 2\frac{x_2^2}{|x|^4}(y_1^2 y_2 + 2y_2^3) - \frac{8}{3}\frac{x_1^3 x_2}{|x|^6}y_1^3$$
$$\left.\left.-8\frac{x_1^2 x_2^2}{|x|^6}y_1^2 y_2 - 8\frac{x_1 x_2^3}{|x|^6}y_1 y_2^2 - \frac{8}{3}\frac{x_2^4}{|x|^6}y_2^3\right]\right\} + O(|x|^{-3}),$$

with class $\mathcal{A}$ component

$$\left((D(x_p,y))_{32}\right)^{\mathcal{A}}$$
$$= a_2\left\{[\mu^2 r(2\ln r + 1) - 4h^2\mu(\lambda+2\mu)r^{-1}\sin\theta\right.$$
$$-2\mu^2 y_2 \ln r - [2\mu^2 y_2 + \mu^2 y_1 \sin(2\theta) - \mu^2 y_2 \cos(2\theta)]$$
$$+r^{-1}\left[\left(\tfrac{1}{2}\mu^2(y_1^2 + 3y_2^2) + 4h^2\mu(\lambda+2\mu)\right)\sin\theta + \mu^2 y_1 y_2 \cos\theta\right.$$

$$-\frac{1}{2}\mu^2(y_1^2 - y_2^2)\sin(3\theta) + \mu^2 y_1 y_2 \cos(3\theta)\Big]$$

$$+r^{-2}\Big[\frac{1}{3}\mu^2(y_1^3 + 3y_1 y_2^2)\sin(2\theta) - \frac{2}{3}\mu^2 y_2^3 \cos(2\theta)$$

$$-\frac{1}{3}\mu^2(y_1^3 - 3y_1 y_2^2)\sin(4\theta) + \frac{1}{3}\mu^2(3y_1^2 y_2 - y_2^3)\cos(4\theta)\Big]\Big\} + O(r^{-3})$$

and residual

$$\big((D(x,y))_{32}\big)^\infty = a_2\Big\{-4h^2\mu(\lambda + 2\mu)\frac{x_2}{x_1^2 + x_2^2} + \mu^2 x_2[1 + \ln(x_1^2 + x_2^2)]\Big\},$$

or, in polar coordinates,

$$\big((D(x_p,y))_{32}\big)^\infty = a_2[\mu^2 r(2\ln r + 1) - 4h^2\mu(\lambda + 2\mu)r^{-1}]\sin\theta.$$

1.17 Remark. These expansions show that the coefficients $m_0, \ldots, m_{12}$ of the far-field pattern (1.15) for $(D^{(2)})^{\mathscr{A}}(x_p, y)$ are

$$m_0 = a_2\mu^2 y_1,$$

$$m_1 = \frac{1}{2}a_2\mu^2 y_2,$$

$$m_2 = \frac{3}{2}a_2\mu^2 y_2,$$

$$m_3 = \frac{1}{2}a_2\mu^2(y_1^2 - y_2^2),$$

$$m_4 = 2a_2[2h^2\mu(\lambda + 2\mu) + \mu^2 y_2^2],$$

$$m_5 = 0,$$

$$m_6 = a_2\mu^2 y_1 y_2,$$

$$m_7 = a_2\Big[4h^2\mu(\lambda + 2\mu)y_1 + \frac{1}{2}\mu^2(y_1^3 + y_1 y_2^2)\Big],$$

$$m_8 = -a_2\Big[4h^2\mu(\lambda + 2\mu)y_2 + \frac{1}{2}\mu^2(y_1^2 y_2 + y_2^3)\Big],$$

$$m_9 = a_2\Big[4h^2\mu(\lambda + 2\mu)y_1 + \frac{1}{6}\mu^2(9y_1 y_2^2 + y_1^3)\Big],$$

$$m_{10} = -a_2\Big[4h^2\mu(\lambda + 2\mu)y_2 - \frac{1}{6}\mu^2(3y_1^2 y_2 - 5y_2^3)\Big],$$

$$m_{11} = \frac{1}{3}\mu^2(y_1^3 + 3y_1 y_2^2),$$

$$m_{12} = -\frac{2}{3}\mu^2 y_2^3.$$

The asymptotic expansion of the first entry of $D^{(3)}$ in Cartesian coordinates is

$$(D(x,y))_{13}$$

$$= a_2\mu^2\Big\{(2\ln|x| + 1)(y_1 - x_1)$$

$$+ 2\Big(\frac{x_1^2}{|x|^2}y_1 + \frac{x_1 x_2}{|x|^2}y_2\Big)$$

$$+2\left[-\frac{1}{2}\frac{x_1}{|x|^2}(3y_1^2+y_2^2)-\frac{x_2}{|x|^2}y_1y_2+\frac{x_1^3}{|x|^4}y_1^2\right.$$
$$\left.+2\frac{x_1^2x_2}{|x|^4}y_1y_2+\frac{x_1x_2}{|x|^4}y_2^2\right]$$
$$+\left[\frac{1}{|x|^2}(y_1^3+y_1y_2^2)-2\frac{x_1^2}{|x|^4}(2y_1^3+y_1y_2^2)-2\frac{x_1x_2}{|x|^4}(3y_1^2y_2+y_2^3)\right.$$
$$-2\frac{x_2}{|x|^4}y_1y_2^2+\frac{8}{3}\frac{x_1^4}{|x|^6}y_1^3+8\frac{x_1^3x_2}{|x|^6}y_1^2y_2$$
$$\left.+8\frac{x_1^2x_2^2}{|x|^6}y_1y_2^2+\frac{8}{3}\frac{x_1x_2^3}{|x|^6}y_2^3\right]$$
$$+\left[\frac{x_1}{|x|^4}\left(\frac{5}{4}y_1^4+\frac{3}{2}y_1^2y_2^2+\frac{1}{4}y_2^4\right)+\frac{x_2}{|x|^4}(y_1^3y_2+y_1y_2^3)\right.$$
$$-\frac{x_1^3}{|x|^6}\left(\frac{10}{3}y_1^4+2y_1^2y_2^2\right)-\frac{x_1^2x_2}{|x|^6}(8y_1^3y_2+4y_1y_2^3)$$
$$-\frac{x_1x_2^2}{|x|^6}(6y_1^2y_2^2+2y_2^4)-\frac{4}{3}\frac{x_2^3}{|x|^6}y_1y_2^3$$
$$+2\frac{x_1^5}{|x|^8}y_1^4+8\frac{x_1^4x_2}{|x|^8}y_1^3y_2+12\frac{x_1^3x_2^2}{|x|^8}y_1^2y_2^2$$
$$\left.\left.+8\frac{x_1^2x_2^2}{|x|^8}y_1y_2^3+2\frac{x_1x_2^4}{|x|^8}y_2^4\right]\right\}+O(|x|^{-4}),$$

with class $\mathscr{A}$ component

$$\left((D(x_p,y))_{13}\right)^{\mathscr{A}}$$
$$=a_2\mu^2\left\{[2y_1(\ln r+1)-r(2\ln r+1)\cos\theta+y_2\sin(2\theta)+y_1\cos(2\theta)]\right.$$
$$+r^{-1}\left[-y_1y_2\sin\theta-\left(\frac{3}{2}y_1^2+\frac{1}{2}y_2^2\right)\cos\theta\right.$$
$$\left.+y_1y_2\sin(3\theta)+\left(\frac{1}{2}y_1^2-\frac{1}{2}y_2^2\right)\cos(3\theta)\right]$$
$$+r^{-2}\left[-\left(y_1^2y_2+\frac{1}{3}y_2^3\right)\sin(2\theta)-\frac{2}{3}y_1^3\cos(2\theta)\right.$$
$$\left.+\left(y_1^2y_2-\frac{1}{3}y_2^3\right)\sin(4\theta)+\left(\frac{1}{3}y_1^3-y_1y_2^2\right)\cos(4\theta)\right]$$
$$+r^{-3}\left[-\left(y_1^3y_2+\frac{1}{3}y_1y_2^3\right)\sin(3\theta)\right.$$
$$+\left(-\frac{5}{12}y_1^4+\frac{1}{2}y_1^2y_2^2+\frac{1}{4}y_2^4\right)\cos(3\theta)$$
$$+(y_1^3y_2-y_1y_2^3)\sin(5\theta)$$
$$\left.\left.+\left(\frac{1}{4}y_1^4-\frac{3}{2}y_1^2y_2^2+\frac{1}{4}y_2^4\right)\cos(5\theta)\right]\right\}+O(r^{-4})$$

and residual

$$\left(\left(D(x,y)\right)_{13}\right)^{\infty} = -a_2\mu^2\left\{\frac{(-x_1^2+x_2^2)y_1-2x_1x_2y_2}{x_1^2+x_2^2}\right.$$
$$\left. +x_1[1+\ln(x_1^2+x_2^2)]-y_1[2+\ln(x_1^2+x_2^2)]\right\},$$

or, in polar coordinates,

$$\left(\left(D(x_p,y)\right)_{13}\right)^{\infty} = a_2\mu^2[2y_1(\ln r+1)-r(2\ln r+1)\cos\theta$$
$$+y_2\sin(2\theta)+y_1\cos(2\theta)].$$

The asymptotic expansion of the second entry of $D^{(3)}$ in Cartesian coordinates is

$$\left(D(x,y)\right)_{23}$$
$$= a_2\mu^2\left\{(2\ln|x|+1)(y_2-x_2)+2\left(\frac{x_1x_2}{|x|^2}y_1+\frac{x_2^2}{|x|^2}y_2\right)\right.$$
$$+\left[-2\frac{x_1}{|x|^2}y_1y_2-2\frac{x_2}{|x|^2}(y_1^2+3y_2^2)\right.$$
$$\left.+2\frac{x_1^2x_2}{|x|^4}y_1^2+4\frac{x_1x_2^2}{|x|^4}y_1y_2+2\frac{x_2^3}{|x|^4}y_2^2\right]$$
$$+\left[\frac{1}{|x|^2}(y_1^2y_2+y_2^3)-2\frac{x_1^2}{|x|^4}y_1^2y_2-2\frac{x_1x_2}{|x|^4}(y_1^3+3y_1y_2^2)\right.$$
$$-2\frac{x_2^2}{|x|^4}(y_1^2y_2+4y_2^3)+\frac{8}{3}\frac{x_1^3x_2}{|x|^6}y_1^3+8\frac{x_1^2x_2^2}{|x|^6}y_1^2y_2$$
$$\left.+8\frac{x_1x_2^3}{|x|^6}y_1y_2^2+\frac{8}{3}\frac{x_2^4}{|x|^6}y_2^3\right]$$
$$+\left[2\frac{x_1}{|x|^4}(y_1^3y_2+y_1y_2^3)+2\frac{x_2}{|x|^4}\left(\frac{1}{2}y_1^4+3y_1^2y_2^2+\frac{5}{2}y_2^4\right)\right.$$
$$-\frac{8}{3}\frac{x_1^3}{|x|^6}y_1^3y_2-4\frac{x_1^2x_2}{|x|^6}(y_1^4+3y_1^2y_2^2)$$
$$-8\frac{x_1x_2^2}{|x|^6}(y_1^3y_2+2y_1y_2^3)-4\frac{x_2^3}{|x|^6}\left(y_1^2y_2^2+\frac{5}{3}y_2^4\right)$$
$$+4\frac{x_1^4}{|x|^8}y_1^4+16\frac{x_1^3x_2}{|x|^8}y_1^3y_2+24\frac{x_1^2x_2^3}{|x|^8}y_1^2y_2^2$$
$$\left.\left.+16\frac{x_1x_2^4}{|x|^8}y_1y_2^3+4\frac{x_2^5}{|x|^8}y_2^4\right]\right\}+O(|x|^{-4}),$$

with class $\mathscr{A}$ component

$$\left(\left(D(x_p,y)\right)_{23}\right)^{\mathscr{A}}$$
$$= a_2\mu^2\left\{[2y_2(\ln r+1)-r(2\ln r+1)\sin\theta+y_1\sin(2\theta)-y_2\cos(2\theta)]\right.$$

$$+r^{-1}\Big[-\Big(\tfrac{1}{2}y_1^2+\tfrac{3}{2}y_2^2\Big)\sin\theta-\eta_1 y_2\cos\theta$$
$$+\Big(\tfrac{1}{2}y_1^2-\tfrac{1}{2}y_2^2\Big)\sin(3\theta)-y_1 y_2\cos(3\theta)\Big]$$
$$+r^{-2}\Big[-\Big(\tfrac{1}{3}y_1^3+y_1 y_2^2\Big)\sin(2\theta)+\tfrac{2}{3}y_2^3\cos(2\theta)$$
$$+\Big(\tfrac{1}{3}y_1^3-y_1 y_2^2\Big)\sin(4\theta)+\Big(\tfrac{1}{3}y_2^3-y_1^2 y_2\Big)\cos(4\theta)\Big]$$
$$+r^{-3}\Big[\Big(-\tfrac{1}{4}y_1^4-\tfrac{1}{2}y_1^2 y_2^2+\tfrac{5}{12}y_2^4\Big)\sin(3\theta)$$
$$+\Big(\tfrac{1}{3}y_1^3 y_2+y_1 y_2^3\Big)\cos(3\theta)$$
$$+\Big(\tfrac{1}{4}y_1^4-\tfrac{3}{2}y_1^2 y_2^2+\tfrac{1}{4}y_2^4\Big)\sin(5\theta)$$
$$+(y_1 y_2^3-y_1^3 y_2)\cos(5\theta)\Big]\Big\}+O(r^{-4})$$

and residual

$$\big((D(x,y))_{23}\big)^\infty=-a_2\mu^2\bigg\{\frac{(x_1^2-x_2^2)y_2-2x_1 x_2 y_1}{x_1^2+x_2^2}$$
$$+x_2[1+\ln(x_1^2+x_2^2)]-y_2[2+\ln(x_1^2+x_2^2)]\bigg\},$$

or, in polar coordinates,

$$\big((D(x_p,y))_{23}\big)^\infty=a_2\mu^2[2y_2(\ln r+1)-r(2\ln r+1)\sin\theta$$
$$+y_1\sin(2\theta)-y_2\cos(2\theta)].$$

The asymptotic expansion of the third entry of $D^{(3)}$ in Cartesian coordinates is

$$(D(x,y))_{33}$$

$$=a_2\bigg\{\mu^2|x|^2\ln|x|-2\mu^2[(x_1\ln|x|)y_1+(x_2\ln|x|)y_2]-\mu^2(x_1 y_1+x_2 y_2)$$

$$+\mu^2(\ln|x|)(y_1^2+y_2^2)-4h^2\mu(\lambda+2\mu)\ln|x|-4h^2\mu(\lambda+3\mu)$$

$$+\mu^2\bigg[\tfrac{1}{2}(y_1^2+y_2^2)+\frac{x_1^2}{|x|^2}y_1^2+2\frac{x_1 x_2}{|x|^2}y_1 y_1+\frac{x_2^2}{|x|^2}y_2^2\bigg]$$

$$+4h^2\mu(\lambda+2\mu)\bigg(\frac{x_1}{|x|^2}y_1+\frac{x_2}{|x|^2}y_2\bigg)$$

$$+\mu^2\bigg[-\frac{x_1}{|x|^2}(y_1^3+y_1 y_2^2)-\frac{x_2}{|x|^2}(y_1^2 y_2+y_2^3)+\frac{2}{3}\frac{x_1^3}{|x|^4}y_1^3$$

$$+2\frac{x_1^2 x_2}{|x|^4}y_1^2 y_2+2\frac{x_1 x_2^2}{|x|^4}y_1 y_2^2+\frac{2}{3}x_2^3|x|^4 y_2^3\bigg]$$

$$-2h^2\mu(\lambda+2\mu)\bigg[\frac{1}{|x|^2}(y_1^2+y_2^2)-2\frac{x_1^2}{|x|^4}y_1^2-4\frac{x_1 x_2}{|x|^4}y_1 y_2-2\frac{x_2^2}{|x|^4}y_2^2\bigg]$$

$$+ \mu^2 \left[\frac{1}{4} \frac{1}{|x|^2} (y_1^2 + y_2^2)^2 - \frac{x_1^2}{|x|^4} (y_1^4 + y_1^2 y_2^2) \right.$$

$$- 2 \frac{x_1 x_2}{|x|^4} (y_1^3 y_2 + y_1 y_2^3) - \frac{x_2^2}{|x|^4} (y_1^2 y_2^2 + y_2^4)$$

$$+ \frac{2}{3} \frac{x_1^4}{|x|^6} y_1^4 + \frac{8}{3} \frac{x_1^3 x_2}{|x|^6} y_1^3 y_2 + 4 \frac{x_1^2 x_2^2}{|x|^6} y_1^2 y_2^2$$

$$\left. + \frac{8}{3} \frac{x_1 x_2^3}{|x|^6} y_1 y_2^3 + \frac{2}{3} \frac{x_2^4}{|x|^6} y_2^4 \right] \right\} + O(|x|^{-3}),$$

with class $\mathscr{A}$ component

$$\left((D(x_p, y))_{33} \right)^{\mathscr{A}}$$

$$= a_2 \left\{ \mu^2 r^2 \ln r - 4h^2 \mu (\lambda + 2\mu) \ln r - 4h^2 \mu (\lambda + 3\mu) \right.$$

$$+ [-\mu^2 r (2 \ln r + 1) + 4h^2 \mu (\lambda + 2\mu) r^{-1}] (y_1 \cos\theta + y_2 \sin\theta)$$

$$+ \mu^2 \left[(y_1^2 + y_2^2)(\ln r + 1) + y_1 y_2 \sin(2\theta) + \tfrac{1}{2} (y_1^2 - y_2^2) \cos(2\theta) \right]$$

$$+ r^{-1} \left[-\tfrac{1}{2} \mu^2 (y_1^2 y_2 + y_2^3) \sin\theta - \tfrac{1}{2} \mu^2 (y_1^3 + y_1 y_2^2) \cos\theta \right.$$

$$\left. + \mu^2 \left[\left(\tfrac{1}{2} y_1^2 y_2 - \tfrac{1}{6} y_2^3 \right) \sin(3\theta) + \mu^2 \left(\tfrac{1}{6} y_1^3 - \tfrac{1}{2} y_1 y_2^2 \right) \cos(3\theta) \right] \right.$$

$$+ r^{-2} \left[\left(4h^2 \mu (\lambda + 2\mu) y_1 y_2 - \mu^2 \left(\tfrac{1}{3} y_1^3 y_2 + \tfrac{1}{3} y_1 y_2^3 \right) \right) \sin(2\theta) \right.$$

$$+ \left(2h^2 \mu (\lambda + 2\mu)(y_1^2 - y_2^2) - \mu^2 \left(\tfrac{1}{6} y_1^4 - \tfrac{1}{6} y_2^4 \right) \right) \cos(2\theta)$$

$$+ \mu^2 \left(\tfrac{1}{3} y_1^3 y_2 - \tfrac{1}{3} y_1 y_2^3 \right) \sin(4\theta)$$

$$\left. \left. + \mu^2 \left(\tfrac{1}{12} y_1^4 - \tfrac{1}{2} y_1^2 y_2^2 + \tfrac{1}{12} y_2^4 \right) \cos(4\theta) \right] \right\} + O(r^{-3})$$

and residual

$$\left((D(x, y))_{33} \right)^{\infty} = a_2 \left\{ -4h^2 \mu (\lambda + 3\mu) + 4h^2 \mu (\lambda + 2\mu) \frac{x_1 y_1 + x_2 y_2}{x_1^2 + x_2^2} \right.$$

$$+ \tfrac{1}{2} \mu^2 (x_1^2 + x_2^2) \ln(x_1^2 + x_2^2) - 2h^2 \mu (\lambda + 2\mu) \ln(x_1^2 + x_2^2)$$

$$\left. - \mu^2 (x_1 y_1 + x_2 y_2)[1 + \ln(x_1^2 + x_2^2)] \right\},$$

or, in polar coordinates,

$$\left((D(x_p, y))_{33} \right)^{\infty} = a_2 \{ \mu^2 r^2 \ln r - 4h^2 \mu (\lambda + 2\mu) \ln r - 4h^2 \mu (\lambda + 3\mu)$$

$$+ [-\mu^2 r (2 \ln r + 1) + 4h^2 \mu (\lambda + 2\mu) r^{-1}]$$

$$\times (y_1 \cos\theta + y_2 \sin\theta) \}.$$

1.18 Remark. These expansions show that the coefficients $m_0, \ldots, m_{12}$ of the far-field pattern (1.15) for $(D^{(3)})^{\mathscr{A}}(x_p, y)$ are

$$m_0 = -a_2 \mu^2 y_1 y_2,$$

$$m_1 = -a_2 \mu^2 \left(\tfrac{3}{4} y_1^2 + \tfrac{1}{4} y_2^2 \right),$$

$$m_2 = -a_2 \mu^2 \left(\tfrac{1}{4} y_1^2 + \tfrac{3}{4} y_2^2 \right),$$

$$m_3 = a_2 \mu^2 \left(\tfrac{1}{6} y_2^3 - \tfrac{1}{2} y_1^2 y_2 \right),$$

$$m_4 = -\tfrac{2}{3} a_2 \mu^2 y_2^3,$$

$$m_5 = -\tfrac{2}{3} a_2 \mu^2 y_1^3,$$

$$m_6 = \alpha_2 \mu^2 \left(\tfrac{1}{6} y_1^3 - \tfrac{1}{2} y_1 y_2^2 \right),$$

$$m_7 = -a_2 \mu^2 \left(\tfrac{1}{2} y_1^3 y_2 + \tfrac{1}{6} y_1 y_2^3 \right),$$

$$m_8 = a_2 \mu^2 \left(-\tfrac{5}{24} y_1^4 + \tfrac{1}{4} y_1^2 y_2^2 + \tfrac{1}{8} y_2^4 \right),$$

$$m_9 = -a_2 \mu^2 \left(\tfrac{1}{6} y_1^3 y_2 + \tfrac{1}{2} y_1 y_2^3 \right),$$

$$m_{10} = a_2 \mu^2 \left(-\tfrac{1}{8} y_1^4 - \tfrac{1}{4} y_1^2 y_2^2 + \tfrac{5}{24} y_2^4 \right),$$

$$m_{11} = 4h^2 \mu (\lambda + 2\mu) y_1 y_2 - \tfrac{1}{3} \mu^2 (y_1^3 y_2 + y_1 y_2^3),$$

$$m_{12} = 2h^2 \mu (\lambda + 2\mu)(y_1^2 - y_2^2) - \tfrac{1}{6} \mu^2 (y_1^4 - y_2^4).$$

1.19 Remark. In what follows, we need certain properties of the matrix D and its class $\mathscr{A}$ component, which we will establish using positional notation (by means of nested sets of parentheses of various sizes) to specify the correct order of the operations performed on the 'basic' matrix $D(x, y)$. We recall (see Remark 1.13) that $\mathscr{A}$-expansion is always executed with respect to x. For example,

$\left((D(x, y))^{\mathscr{A}} \right)^{\mathsf{T}}$: $\mathscr{A}$-expansion, followed by transposition;

$\left((D(y, x))^{\mathsf{T}} \right)^{\mathscr{A}}$: the interchange of x and y, followed by transposition, followed by $\mathscr{A}$-expansion;

$\left(D^{\mathscr{A}}(y, x) \right)^{\mathsf{T}}$: $\mathscr{A}$-expansion, followed by the interchange of x and y, followed by transposition;

and so on. When $\mathscr{A}$-expansion is involved, it is essential to indicate the precise sequence of operations in a chain as two consecutive steps may not commute. This makes the notation somewhat cumbersome, but it eliminates misunderstanding and errors.

As explained in Remark 1.15, the matrix $(D(x, y))^{\mathscr{A}}$ is created from the class $\mathscr{A}$ components $\left((D(x, y))^{(j)} \right)^{\mathscr{A}}$ of the columns $(D(x, y))^{(j)}$ of matrix $D(x, y)$, with the

$\mathscr{A}$-expansions involved in its definition carried out individually on each entry. Thus, from (1.9) it follows that

$$\left(D(x,y)\right)^{\mathscr{A}} = \left(\left(D(y,x)\right)^{\mathsf{T}}\right)^{\mathscr{A}}, \qquad (1.18)$$

so, by (1.17),

$$\left(\left(D(x,y)\right)^{\mathscr{A}}\right)_{ij} = \left(\left(D(x,y)\right)_{ij}\right)^{\mathscr{A}}$$
$$= \left(\left(D(y,x)\right)_{ji}\right)^{\mathscr{A}} = \left(\left(D(y,x)\right)^{\mathscr{A}}\right)_{ji}; \qquad (1.19)$$

hence, by (1.19) and (1.18),

$$\left(D(x,y)\right)^{\mathscr{A}} = \left(\left(D(y,x)\right)^{\mathscr{A}}\right)^{\mathsf{T}} = \left(\left(D(y,x)\right)^{\mathsf{T}}\right)^{\mathscr{A}}, \qquad (1.20)$$

where the middle term indicates the interchange of x and y in the 'basic' matrix $D(x,y)$, followed by $\mathscr{A}$-expansion, followed by transposition, and the right-hand side term has the last two operations performed in reverse order.

We mention that the following sequence of operations *does not* work:

$$\left(D(x,y)\right)^{\mathscr{A}} = \left(\left(D(y,x)\right)^{\mathsf{T}}\right)^{\mathscr{A}} \neq \left(\left(D^{\mathscr{A}}(y,x)\right)\right)^{\mathsf{T}}.$$

In other words, $\left(D(x,y)\right)^{\mathscr{A}}$ is not the same matrix as the one generated by the $\mathscr{A}$-expansion of $D(x,y)$, followed by the interchange of the points x and y, followed by transposition. Differently put, the matrix $\left(D(x,y)\right)^{\mathscr{A}}$ does not have property (1.9).

1.2.2 Asymptotic Expansion of Matrix $P(x,y)$

We have

$$P(x,y) = \left(T(\partial_y)D(y,x)\right)^{\mathsf{T}} = \left(T(\partial_y)\left(D(x,y)\right)^{\mathsf{T}}\right)^{\mathsf{T}},$$

so

$$\left(P(x,y)\right)_{ij} = \left(T(\partial_y)\left(D(x,y)\right)^{\mathsf{T}}\right)_{ji} = \sum_{h=1}^{3} T_{jh}(\partial_y)\left(\left(D(x,y)\right)^{\mathsf{T}}\right)_{hi}$$
$$= \sum_{h=1}^{3} T_{jh}(\partial_y)\left(D(x,y)\right)_{ih}, \quad i,j = 1,2,3;$$

in other words,

$$\left(P(x,y)\right)^{(j)} = \sum_{h=1}^{3} T_{jh}(\partial_y)\left(D(x,y)\right)^{(h)}, \quad j = 1,2,3. \qquad (1.21)$$

Since the action of the linear operator $T(\partial_y)$ does not interfere with the structure of the columns of $D(x,y)$ related to class $\mathscr{A}$, which are computed in terms of x, this shows that the columns of P have a similar class $\mathscr{A}$ structure; consequently, $P(x,y)$ consists of a class $\mathscr{A}$ component $P^{\mathscr{A}}$ (constructed from (1.21) with $D^{\mathscr{A}}$) and a residual P^{∞} (constructed from (1.21) with D^{∞}). As it turns out, direct calculation shows that $P^{\infty} = 0$; hence,

$$P(x,y) = \left(P(x,y)\right)^{\mathscr{A}}.$$

The far-field expansions of the components of $P^{(1)}$ in polar coordinates are

$$
\begin{aligned}
\left(P(x_p,y)\right)_{11} \\
= a_2 \{ r^{-1} [& 2h^2\mu^3 n_2 \sin\theta + 2h^2\mu^2(2\lambda+3\mu)n_1\cos\theta \\
& - 2h^2\mu^3 n_2\sin(3\theta) - 2h^2\mu^3 n_1\cos(3\theta)] \\
+ r^{-2}[& 4h^2\mu^2(\mu n_2 y_1 + (\lambda+\mu)n_1 y_2)\sin(2\theta) \\
& + 4h^2\mu^2(\lambda+2\mu)n_1 y_1\cos(2\theta) \\
& - 4h^2\mu^3(n_2 y_1 + n_1 y_2)\sin(4\theta) \\
& - 4h^2\mu^3(n_1 y_1 - n_2 y_2)\cos(4\theta)] \\
+ r^{-3}[& 2h^2\mu^2(8h^2(\lambda+2\mu)n_2 + \mu n_2(3y_1^2 + y_2^2) \\
& + 2(2\lambda+3\mu)n_1 y_1 y_2)\sin(3\theta) \\
& + 2h^2\mu^2(8h^2(\lambda+2\mu)n_1 + (2\lambda+5\mu)n_1 y_1^2 \\
& \qquad - (2\lambda+\mu)n_1 y_2^2 - 2\mu n_2 y_1 y_2)\cos(3\theta) \\
& - 6h^2\mu^3(n_2(y_1^2 - y_2^2) + 2n_1 y_1 y_2)\sin(5\theta) \\
& + 6h^2\mu^3(n_1(y_2^2 - y_1^2) + 2n_2 y_1 y_1)\cos(5\theta)]\} + O(r^{-4}),
\end{aligned}
$$

$$
\begin{aligned}
\left(P(x_p,y)\right)_{21} \\
= a_2 \{ r^{-1} [& 2h^2\mu^2(2\lambda+\mu)n_1\sin\theta + 2h^2\mu^3 n_2\cos\theta \\
& - 2h^2\mu^3 n_1\sin(3\theta) + 2h^2\mu^3 n_2\cos(3\theta)] \\
+ r^{-2}[& 4h^2\mu^2(\mu n_2 y_2 + (\lambda+\mu)n_1 y_1)\sin(2\theta) \\
& - 4h^2\lambda\mu^2 n_1 y_2\cos(2\theta) \\
& - 4h^2 m^3(n_1 y_1 - n_2 y_2)\sin(4\theta)
\end{aligned}
$$

$$+4h^2\mu^3(n_2y_1+n_1y_2)\cos(4\theta)]$$
$$+r^{-3}[2h^2\mu^2(8h^2(\lambda+2\mu)n_1+(2\lambda+3\mu)n_1y_1^2$$
$$-(2\lambda-\mu)n_1y_2^2+2\mu n_2y_1y_2)\sin(3\theta)$$
$$-2h^2\mu^2(8h^2(\lambda+2\mu)n_2+\mu n_2(y_1^2+3y_2^2)$$
$$+2(2\lambda+\mu)n_1y_1y_2)\cos(3\theta)$$
$$+6h^2\mu^3(n_1(y_2^2-y_1^2)+2n_2y_1y_2)\sin(5\theta)$$
$$+6h^2\mu^3(n_2(y_1^2-y_2^2)+2n_1y_1y_2)\cos(5\theta)]\}+O(r^{-4}),$$

$$\left(P(x_p,y)\right)_{31}$$
$$=-a_2\{4h^2\mu^2(\lambda+\mu)n_1(\ln r+1)+2h^2\mu^3 n_2\sin(2\theta)+2h^2\mu^3 n_1\cos(2\theta)$$
$$+r^{-1}[2h^2\mu^2(\mu n_2y_1+(2\lambda+\mu)n_1y_2)\sin\theta$$
$$+2h^2\mu^2((2\lambda+3\mu)n_1y_1+\mu n_2y_2)\cos\theta$$
$$-3h^2\mu^3(n_2y_1+n_1y_2)\sin(3\theta)$$
$$-2h^2\mu^3(n_1y_1-n_2y_2)\cos(3\theta)]$$
$$+r^{-2}[2h^2\mu^2(\mu n_2(y_1^2+y_2^2)+2(\lambda+\mu)n_1y_1y_2)\sin(2\theta)$$
$$+2h^2\mu^2((\lambda+2\mu)n_1y_1^2-\lambda n_1y_2^2)\cos(2\theta)$$
$$-2h^2\mu^3(n_2(y_1^2-y_2^2)+2n_1y_1y_2)\sin(4\theta)$$
$$+2h^2\mu^3(n_1(y_2^2-y_1^2)+2n_2y_1y_2)\cos(4\theta)]\}+O(r^{-3}).$$

1.20 Remark. These expansions show that the coefficients $m_0,\ldots,m_{12}$ of the far-field pattern (1.15) for $\left(P(x_p,y)\right)^{(1)}$ are

$$m_0=2a_2h^2n_2\mu^3,$$
$$m_1=a_2h^2n_1\mu^2(2\lambda+3\mu),$$
$$m_2=a_2h^2n_1\mu^2(2\lambda+\mu),$$
$$m_3=2a_2h^2(n_2y_1+n_1y_2)\mu^3,$$
$$m_4=4a_2h^2n_1y_2\lambda\mu^2,$$
$$m_5=4a_2h^2n_1y_1\mu^2(\lambda+2\mu),$$
$$m_6=2a_2h^2(-n_1y_1+n_2y_2)\mu^3,$$
$$m_7=a_2h^2\mu^2[n_2(3y_1^2+y_2^2)\mu+8h^2n_2(\lambda+2\mu)+2n_1y_1y_2(2\lambda+3\mu)],$$

$$m_8 = a_2 h^2 \mu^2 \big[-2n_2 y_1 y_2 \mu + 8h^2 n_1 (\lambda + 2\mu)$$
$$+ n_1 \big(-y_2^2 (2\lambda + \mu) + y_1^2 (2\lambda + 5\mu) \big) \big],$$

$$m_9 = a_2 h^2 \mu^2 [n_2 (y_1^2 + 3y_2^2)\mu + 2n_1 y_1 y_2 (2\lambda + \mu) + 8h^2 n_2 (\lambda + 2\mu)],$$

$$m_{10} = a_2 h^2 \mu^2 \big[2n_2 y_1 y_2 \mu + 8h^2 n_1 (\lambda + 2\mu)$$
$$+ n_1 \big(y_2^2 (-2\lambda + \mu) + y_1^2 (2\lambda + 3\mu) \big) \big],$$

$$m_{11} = 2a_2 h^2 \mu^2 [n_2 (y_1^2 + y_2^2)\mu + 2n_1 y_1 y_2 (\lambda + \mu)],$$

$$m_{12} = 2a_2 h^2 n_1 \mu^2 [-y_2^2 \lambda + y_1^2 (\lambda + 2\mu)].$$

The far-field expansions of the components of $P^{(2)}$ in polar coordinates are

$$(P(x_p, y))_{12}$$
$$= a_2 \{ r^{-1} [2h^2 \mu^3 n_1 \sin\theta + 2h^2 \mu^2 (2\lambda + \mu) n_2 \cos\theta$$
$$- 2h^2 \mu^3 n_1 \sin(3\theta) + 2h^2 \mu^3 n_2 \cos(3\theta)]$$
$$+ r^{-2} [4h^2 \mu^2 (\mu n_1 y_1 + (\lambda + \mu) n_2 y_2) \sin(2\theta)$$
$$+ 4h^2 \lambda \mu^2 n_2 y_1 \cos(2\theta)$$
$$+ 4h^2 \mu^3 (n_2 y_2 - n_1 y_1) \sin(4\theta)$$
$$+ 4h^2 \mu^3 (n_1 y_1 + n_1 y_2) \cos(4\theta)]$$
$$+ r^{-3} [2h^2 \mu^2 (8h^2 (\lambda + 2\mu) n_1 + \mu n_1 (3y_1^2 + y_2^2)$$
$$+ 2(2\lambda + \mu) n_2 y_1 y_2) \sin(3\theta)$$
$$- 2h^2 \mu^2 (8h^2 (\lambda + 2\mu) n_2 - (2\lambda - \mu) n_2 y_1^2$$
$$+ (2\lambda + 3\mu) n_2 y_2^2 + 2\mu n_1 y_1 y_2) \cos(3\theta)$$
$$+ 6h^2 \mu^3 (n_1 (y_2^2 - y_1^2) + 2n_2 y_1 y_2) \sin(5\theta)$$
$$+ 6h^2 \mu^3 (n_2 (y_1^2 - y_2^2) + 2n_1 y_1 y_2) \cos(5\theta)] \} + O(r^{-4}),$$

$$(P(x_p, y))_{22}$$
$$= a_2 \{ r^{-1} [2h^2 \mu^2 (2\lambda + 3\mu) n_2 \sin\theta + 2h^2 \mu^3 n_1 \cos\theta$$
$$+ 2h^2 \mu^3 n_2 \sin(3\theta) + 2h^2 \mu^3 n_1 \cos(3\theta)]$$
$$+ r^{-2} [4h^2 \mu^2 ((\lambda + \mu) n_2 y_1 + \mu n_1 y_2) \sin(2\theta)$$
$$- 4h^2 \mu^2 (\lambda + 2\mu) n_2 y_2 \cos(2\theta)$$
$$+ 4h^2 \mu^3 (n_2 y_1 + n_1 y_2) \sin(4\theta)$$

$$+4h^2\mu^3(n_1y_1-n_2y_2)\cos(4\theta)]$$
$$+r^{-3}[2h^2\mu^2(-8h^2(\lambda+2\mu)n_2+(2\lambda+\mu)n_2y_1^2$$
$$-(2\lambda+5\mu)n_2y_2^2+2\mu n_1y_1y_2)\sin(3\theta)$$
$$-2h^2\mu^2(8h^2(\lambda+2\mu)n_1+\mu n_1(y_1^2+3y_2^2)n_1$$
$$+2(2\lambda+3\mu)n_2y_1y_2)\cos(3\theta)$$
$$+6h^2\mu^3(n_2(y_1^2-y_2^2)+2n_1y_1y_2)\sin(5\theta)$$
$$-6h^2\mu^3(n_1(y_2^2-y_1^2)+2n_2y_1y_2)\cos(5\theta)]\}+O(r^{-4}),$$

$$\left(P(x_p,y)\right)_{32}$$
$$=a_2\{-4h^2\mu^2(\lambda+\mu)n_2(\ln r+1)-2h^2\mu^3n_1\sin(2\theta)+2h^2\mu^3n_2\cos(2\theta)$$
$$+r^{-1}[2h^2\mu^2(\mu n_1y_1+(2\lambda+3\mu)n_2y_2)\sin\theta$$
$$+2h^2\mu^2((2\lambda+\mu)n_2y_1+\mu n_1y_2)\cos\theta$$
$$-2h^2\mu^3(n_1y_1-n_2y_2)\sin(3\theta)$$
$$+2h^2\mu^3(n_2y_1+n_1y_2)\cos(3\theta)]$$
$$+r^{-2}[2h^2\mu^2(\mu n_1(y_1^2+y_2^2)+2(\lambda+\mu)n_2y_1y_2)\sin(2\theta)$$
$$+2h^2\mu^2(\lambda n_2y_1^2-(\lambda+2\mu)n_2y_2^2)\cos(2\theta)$$
$$+2h^2\mu^3(n_1(y_2^2-n_1^2)+2n_2y_1y_2)\sin(4\theta)$$
$$+2h^2\mu^3(n_2(y_1^2-y_2^2)+2n_1y_1y_2)\cos(4\theta)]+O(r^{-3}).$$

1.21 Remark. These expansions show that the coefficients $m_0,\ldots,m_{12}$ of the far-field pattern (1.15) for $\left(P(x_p,y)\right)^{(2)}$ are

$$m_0=2a_2h^2n_1\mu^3,$$
$$m_1=a_2h^2n_2\mu^2(2\lambda+\mu),$$
$$m_2=a_2h^2n_2\mu^2(2\lambda+3\mu),$$
$$m_3=2a_2h^2(n_1y_1-n_2y_2)\mu^3,$$
$$m_4=4a_2h^2n_2y_2\mu^2(\lambda+2\mu),$$
$$m_5=4a_2h^2n_2y_1\lambda\mu^2,$$
$$m_6=2a_2h^2(n_2y_1+n_1y_2)\mu^3,$$
$$m_7=a_2h^2\mu^2[n_1(3y_1^2+y_2^2)\mu+2n_2y_1y_2(2\lambda+\mu)+8h^2n_1(\lambda+2\mu)],$$

$$m_8 = -a_2 h^2 \mu^2 \left[2n_1 y_1 y_2 \mu + 8h^2 n_2 (\lambda + 2\mu) \right.$$
$$\left. + n_2 \left(y_1^2 (-2\lambda + \mu) + y_2^2 (2\lambda + 3\mu) \right) \right],$$
$$m_9 = a_2 h^2 \mu^2 [n_1 (y_1^2 + 3y_2^2)\mu + 8h^2 n_1 (\lambda + 2\mu) + 2n_2 y_1 y_2 (2\lambda + 3\mu)],$$
$$m_{10} = a_2 h^2 \mu^2 \left[2n_1 y_1 y_2 \mu - 8h^2 n_2 (\lambda + 2\mu) \right.$$
$$\left. + n_2 \left(y_1^2 (2\lambda + \mu) - y_2^2 (2\lambda + 5\mu) \right) \right],$$
$$m_{11} = 2a_2 h^2 \mu^2 [n_1 (y_1^2 + y_2^2)\mu + 2n_2 y_1 y_2 (\lambda + \mu)],$$
$$m_{12} = -2a_2 h^2 n_2 \mu^2 [-y_1^2 \lambda + y_2^2 (\lambda + 2\mu)].$$

The far-field expansions of the components of $P^{(3)}$ in polar coordinates are

$$\left(P(x_p, y) \right)_{13}$$
$$= a_2 \{ r^{-2} [4h^2 \mu^2 (\lambda + 2\mu) n_2 \sin(2\theta)$$
$$+ 4h^2 \mu^2 (\lambda + 2\mu) n_1 \cos(2\theta)]$$
$$+ r^{-3} [8h^2 \mu^2 (\lambda + 2\mu)(n_2 y_1 + n_1 y_2) \sin(3\theta)$$
$$+ 8h^2 \mu^2 (\lambda + 2\mu)(n_1 y_1 - n_2 y_2) \cos(3\theta)] \},$$
$$\left(P(x_p, y) \right)_{23}$$
$$= a_2 \{ r^{-2} [4h^2 \mu^2 (\lambda + 2\mu) n_1 \sin(2\theta)$$
$$- 4h^2 \mu^2 (\lambda + 2\mu) n_2 \cos(2\theta)]$$
$$+ r^{-3} [8h^2 \mu^2 (\lambda + 2\mu)(n_1 y_1 - n_2 y_2) \sin(3\theta)$$
$$- 8h^2 \mu^2 (\lambda + 2\mu)(n_2 y_1 + n_1 y_2) \cos(3\theta)] \},$$
$$\left(P(x_p, y) \right)_{33}$$
$$= a_2 \{ r^{-1} [4h^2 \mu^2 (\lambda + 2\mu) n_2 \sin\theta$$
$$+ 4h^2 \mu^2 (\lambda + 2\mu) n_1 \cos\theta]$$
$$+ r^{-2} [4h^2 \mu^2 (\lambda + 2\mu)(n_2 y_1 + n_1 y_2) \sin(2\theta)$$
$$+ 4h^2 \mu^2 (\lambda + 2\mu)(n_1 y_1 - n_2 y_2) \cos(2\theta)] \}.$$

1.22 Remark. These expansions show that the coefficients $m_0, \ldots, m_{12}$ of the far-field pattern (1.15) for $\left(P(x_p, y) \right)^{(3)}$ are

$$m_0 = 0,$$
$$m_1 = 0,$$
$$m_2 = 0,$$
$$m_3 = 0,$$

$$m_4 = 4a_2h^2n_2\mu^2(\lambda+2\mu),$$
$$m_5 = 4a_2h^2n_1\mu^2(\lambda+2\mu),$$
$$m_6 = 0,$$
$$m_7 = 4a_2h^2(n_2y_1+n_1y_2)\mu^2(\lambda+2\mu),$$
$$m_8 = 4a_2h^2(n_1y_1-n_2y_2)\mu^2(\lambda+2\mu),$$
$$m_9 = 4a_2h^2(n_2y_1+n_1y_2)\mu^2(\lambda+2\mu),$$
$$m_{10} = 4a_2h^2(n_1y_1-n_2y_2)\mu^2(\lambda+2\mu),$$
$$m_{11} = 4a_2h^2(n_2y_1+n_1y_2)\mu^2(\lambda+2\mu),$$
$$m_{12} = 4a_2h^2(n_1y_1-n_2y_2)\mu^2(\lambda+2\mu).$$

1.2.3 Components of Vector $D(x,y)\psi(y)$

We are interested in the far field behavior of the vector function $D(x,y)\psi(y)$, where ψ is a vector function defined on ∂S. Since

$$D(x,y)\psi(y) = \psi_1(y)\left(D(x,y)\right)^{(1)} + \psi_2(y)\left(D(x,y)\right)^{(2)} + \psi_3(y)\left(D(x,y)\right)^{(3)},$$

the expansions of the components of $\left(D(x_p,y)\right)^{\mathscr{A}}\psi(y)$ and $\left(D(x_p,y)\right)^{\infty}\psi(y)$ are linear combinations of those of the columns $\left(\left(D(x_p,y)\right)^{\mathscr{A}}\right)^{(i)}$ and $\left(\left(D(x_p,y)\right)^{\infty}\right)^{(i)}$, respectively, with coefficients $\psi_i(y)$.

The class $\mathscr{A}$ component and residual of the first entry of $D(x,y)\psi(y)$ in polar coordinates are

$$\left(\left(D(x_p,y)\right)^{\mathscr{A}}\psi(y)\right)_1$$
$$= a_2\Big\{r^{-1}\Big\{\mu^2\big(y_2\psi_1+y_1\psi_2-y_1y_2\psi_3\big)\sin\theta$$
$$+ \mu^2\big[3y_1\psi_1+y_2\psi_2-\tfrac{1}{2}\big(3y_1^2+y_2^2\big)\psi_3\big]\cos\theta$$
$$- \mu^2\big(y_2\psi_1+y_1\psi_2-y_1y_2\psi_3\big)\sin(3\theta)$$
$$+ \mu^2\big[-y_1\psi_1+y_2\psi_2+\tfrac{1}{2}\big(y_1^2-y_2^2\big)\psi_3\big]\cos(3\theta)\Big\}$$
$$+ r^{-2}\Big\{\big[2\mu^2y_1y_2\psi_1+(4h^2\mu(\lambda+2\mu)+\mu^2(y_1^2+y_2^2))\psi_2$$
$$- \tfrac{1}{3}\mu^2(3y_1^2y_2+y_2^3)\psi_3\big]\sin(2\theta)$$
$$+ \big[(4h^2\mu(\lambda+2\mu)+2\mu^2y_1^2)\psi_1-\tfrac{2}{3}\mu^2y_1^3\psi_3\big]\cos(2\theta)$$

$$-\mu^2\left[2y_1y_2\psi_1+(y_1^2-y_2^2)\psi_2-\tfrac{1}{3}(3y_1^2y_2-y_2^3)\psi_3\right]\sin(4\theta)$$

$$+\mu^2\left[(y_2^2-y_1^2)\psi_1+2y_1y_2\psi_2+\tfrac{1}{3}(y_1^3-3y_1y_2^2)\psi_3\right]\cos(4\theta)\big\}$$

$$+r^{-3}\Big[\big(8h^2\mu(\lambda+2\mu)y_2+\tfrac{1}{3}\mu^2(9y_1^2y_2+y_2^3)\big)\psi_1$$

$$+\big(8h^2\mu(\lambda+2\mu)y_1+\mu^2(y_1^3+y_1y_2^2)\big)\psi_2$$

$$-\tfrac{1}{3}\mu^2(3y_1^3y_2+y_1y_2^3)\psi_3\Big]\sin(3\theta)$$

$$+\Big[\big(8h^2\mu(\lambda+2\mu)y_1+\tfrac{1}{3}\mu^2(5y_1^3-3y_1y_2^2)\big)\psi_1$$

$$+\big(8h^2\mu(\lambda+2\mu)y_2+\mu^2(y_1^2y_2+y_2^3)\big)\psi_2$$

$$+\tfrac{1}{12}\mu^2(-5y_1^4+6y_1^2y_2^2+3y_2^4)\psi_3\Big]\cos(3\theta)$$

$$+\mu^2\big[(y_2^3-3y_1^2y_2)\psi_1+(3y_1y_2^2-y_1^3)\psi_2$$

$$+(y_1^3y_2-y_1y_2^3)\psi_3\big]\sin(5\theta)$$

$$+\mu^2\big[(3y_1y_2^2-y_1^3)\psi_1+(3y_1^2y_2-y_2^3)\psi_2$$

$$+\tfrac{1}{4}(y_1^4-6y_1^2y_2^2+y_2^4)\psi_3\big]\cos(5\theta)\Big\}+O(r^{-4})$$

and

$$\big((D(x_p,y))^{\infty}\psi(y)\big)_1$$

$$=-a_2\mu^2\{[2(\ln r+1)+\cos(2\theta)]\psi_1-\sin(2\theta)\psi_2$$

$$-[2y_1(\ln r+1)-r(2\ln r+1)\cos\theta+y_2\sin(2\theta)+y_1\cos(2\theta)]\psi_3\}$$

$$=-a_2\mu^2\{[r(2\ln r+1)\cos\theta]\psi_3$$

$$+[2(\ln r+1)+\cos(2\theta)](\psi_1-y_1\psi_3)+\sin(2\theta)(\psi_2-y_2\psi_3)\}.$$

The class $\mathscr{A}$ component and residual of the second entry of $D(x_p,y)\psi(y)$ in polar coordinates are

$$\big((D(x_p,y))^{\mathscr{A}}\psi(y)\big)_2$$

$$=a_2\Big\{r^{-1}\Big\{\mu^2\big[y_1\psi_1+3y_2\psi_2-\tfrac{1}{2}(y_1^2+3y_2^2)\psi_3\big]\sin\theta$$

$$+\mu^2\big(y_2\psi_1+y_1\psi_2-y_1y_2\psi_3\big)\cos\theta$$

$$+\mu^2\big[-y_1\psi_1+y_2\psi_2+\tfrac{1}{2}(y_1^2-y_2^2)\psi_3\big]\sin(3\theta)$$

$$+\mu^2\big(y_2\psi_1+y_1\psi_2-y_1y_1\psi_3\big)\cos(3\theta)\Big\}$$

$$+r^{-2}\Big\{\big[(4h^2\mu(\lambda+2\mu)+\mu^2(y_1^2+y_2^2))\psi_1+2\mu^2y_1y_2\psi_2$$

$$-\tfrac{1}{3}\mu^2(y_1^3+3y_1y_2^2)\psi_3\big]\sin(2\theta)$$

$$+\big[-(4h^2\mu(\lambda+2\mu)+2\mu^2y_2^2)\psi_2+\tfrac{2}{3}\mu^2y_2^3\psi_3\big]\cos(2\theta)$$

$$+ \mu^2 \left[(y_2^2 - y_1^2) \psi_1 + 2y_1 y_2 \psi_2 + \tfrac{1}{3} (y_1^3 - 3y_1 y_2^2) \psi_3 \right] \sin(4\theta)$$

$$+ \mu^2 \left[2y_1 y_2 \psi_1 + (y_1^2 - y_2^2) \psi_2 + \tfrac{1}{3} (y_2^3 - 3y_1^2 y_2) \psi_3 \right] \cos(4\theta) \Big\}$$

$$+ r^{-3} \Big\{ \left[\left(8h^2 \mu (\lambda + 2\mu) y_1 + \mu^2 (y_1^3 + y_1 y_2^2) \right) \psi_1 \right.$$

$$+ \left(- 8h^2 \mu (\lambda + 2\mu) y_2 + \tfrac{1}{3} \mu^2 (3y_1^2 y_2 - 5y_2^3) \right) \psi_2$$

$$- \left. \tfrac{1}{12} \mu^2 (3y_1^4 + 6y_1^2 y_2^2 - 5y_2^4) \psi_3 \right] \sin(3\theta)$$

$$+ \left[- \left(8h^2 \mu (\lambda + 2\mu) y_2 + \mu^2 (y_1^2 y_2 + y_2^3) \right) \psi_1 \right.$$

$$- \left(8h^2 \mu (\lambda + 2\mu) y_1 + \tfrac{1}{3} \mu^2 (y_1^3 + 9y_1 y_2^2) \right) \psi_2$$

$$+ \left. \tfrac{1}{3} \mu^2 (y_1^3 y_2 + 3y_1 y_2^3) \psi_3 \right] \cos(3\theta)$$

$$+ \mu^2 \left[(3y_1 y_2^2 - y_1^3) \psi_1 + (3y_1^2 y_2 - y_2^3) \psi_2 \right.$$

$$+ \left. \tfrac{1}{4} (y_1^4 - 6y_1^2 y_2^2 + y_2^4) \psi_3 \right] \sin(5\theta)$$

$$+ \mu^2 \left[(3y_1^2 y_2 - y_2^3) \psi_1 + (y_1^3 - 3y_1 y_2^2) \psi_2 \right.$$

$$+ \left. (y_1 y_2^3 - y_1^3 y_2) \psi_3 \right] \cos(5\theta) \Big\} + O(r^{-4})$$

and

$$\left((D(x_p, y))^\infty \psi(y) \right)_2$$

$$= -a_2 \mu^2 \{ \sin(2\theta) \psi_1 + [2(\ln r + 1) - \cos(2\theta)] \psi_2$$

$$+ [-2y_2(\ln r + 1) + r(2\ln r + 1) \sin\theta - y_1 \sin(2\theta) + y^2 \cos(2\theta)] \psi_3 \}$$

$$= -a_2 \mu^2 \{ [(2\ln r + 1) \sin\theta] \psi_3 + \sin(2\theta)(\psi_1 - y_1 \psi_3)$$

$$+ [2(\ln r + 1) - \cos(2\theta)](\psi_2 - y_2 \psi_3) \}.$$

The class $\mathscr{A}$ component and residual of the third entry of $D(x_p, y) \psi(y)$ in polar coordinates are

$$\left((D(x_p, y))^{\mathscr{A}} \psi(y) \right)_3$$

$$= a_2 \Big\{ \mu^2 (\ln + 1)(-2y_1 \psi_1 - 2y_2 \psi_2 + (y_1^2 + y_2^2)) \psi_3$$

$$+ \mu^2 (-y_2 \psi_1 - y_1 \psi_2 + y_1 y_2 \psi_3) \sin(2\theta)$$

$$+ \mu^2 \left(-y_1 \psi_1 + y_2 \psi_2 + \tfrac{1}{2} (y_1^2 - y_2^2) \psi_3 \right) \cos(2\theta)$$

$$+ r^{-1} \Big\{ \left[\mu^2 y_1 y_2 \psi_1 + \left(4h^2 \mu (\lambda + 2\mu) + \tfrac{1}{2} \mu^2 (y_1^2 + 3y_2^2) \right) \psi_2 \right.$$

$$- \left. \tfrac{1}{2} \mu^2 (y_1^2 y_2 + y_2^3) \psi_3 \right] \sin\theta$$

$$+ \left[\left(4h^2 \mu (\lambda + 2\mu) + \tfrac{1}{2} \mu^2 (3y_1^2 + y_2^2) \right) \psi_1 + \mu^2 y_1 y_2 \psi_2 \right.$$

$$-\tfrac{1}{2}\mu^2(y_1^3+y_1y_2^2)\psi_3\big]\cos\theta$$

$$+\mu^2\big[-y_1y_2\psi_1-\tfrac{1}{2}(y_1^2-y_2^2)\psi_2+\tfrac{1}{6}(6y_1^2y_2-y_2^3)\psi_3\big]\sin(3\theta)$$

$$+\big[\tfrac{1}{2}(y_2^2-y_1^2)\psi_1+y_1y_2\psi_2+\tfrac{1}{6}(y_1^3-3y_1y_2^2)\psi_3\big]\cos(3\theta)\big\}$$

$$+r^{-2}\Big\{\big[\tfrac{1}{3}\mu^2(3y_1^2y_2+y_2^3)\psi_1+\tfrac{1}{3}\mu^2(y_1^3+y_1y_2^2)\psi_2$$

$$+\big(4h^2\mu(\lambda+2\mu)y_1y_2-\tfrac{1}{3}\mu^2(y_1^3y_2+y_1y_2^3)\big)\psi_3\big]\sin(2\theta)$$

$$+\big[\tfrac{2}{3}\mu^2(y_1^3\psi_1-y_2^3\psi_2)$$

$$+\big(2h^2\mu(\lambda+2\mu)(y_1^2-y_2^2)-\tfrac{1}{6}\mu^2(y_1^4-y_2^4)\big)\psi_3\big]\cos(2\theta)$$

$$+\mu^2\big[\tfrac{1}{3}(y_2^3-3y_1^2y_2)\psi_1+\tfrac{1}{3}(y_1y_2^2-y_1^3)\psi_2$$

$$+\tfrac{1}{3}(y_1^3y_2-y_1y_2^3)\psi_3\big]\sin(4\theta)$$

$$+\mu^2\big[\tfrac{1}{3}(3y_1y_2^2-y_1^3)\psi_1+\tfrac{1}{3}(3y_1^2y_2-y_2^3)\psi_2$$

$$+\tfrac{1}{12}(y_1^4-6y_1^2y_2^2+y_2^4)\psi_3\big]\cos(4\theta)\Big\}\Big\}+O(r^{-3})$$

and

$$\big((D(x_p,y))^\infty\psi(y)\big)_3$$

$$=a_2\{[\mu^2r(2\ln r+1)-4h^2\mu(\lambda+2\mu)r^{-1}][(\cos\theta)\psi_1+(\sin\theta)\psi_2]$$

$$+\mu^2r^2\ln r-4h^2\mu(\lambda+2\mu)\ln r-4h^2\mu(\lambda+3\mu)$$

$$+[-\mu^2r(2\ln r+1)+4h^2\mu(\lambda+2\mu)r^{-1}](y_2\sin\theta+y_1\cos\theta)\psi_3\}$$

$$=a_2\{[\mu^2r^2\ln r-4h^2\mu(\lambda+2\mu)\ln r-4h^2\mu(\lambda+3\mu)]\psi_3$$

$$+[\mu^2r(2\ln r+1)-4h^2\mu(\lambda+2\mu)r^{-1}]$$

$$\times[(\cos\theta)(\psi_1-y_1\psi_3)+(\sin\theta)(\psi_2-y_2\psi_3)]\}.$$

1.23 Remark. These expansions show that the class $\mathscr{A}$ pattern coefficients for $(D(x_p,y))^{\mathscr{A}}\psi(y)$ are

$$m_0=a_2\mu^2(y_2\psi_1+y_1\psi_2-y_1y_2\psi_3),$$

$$m_1=a_2\mu^2\big[\tfrac{3}{2}y_1\psi_1+\tfrac{1}{2}y_2\psi_2-\tfrac{1}{4}(3y_1^2+y_2^2)\psi_3\big],$$

$$m_2=a_2\mu^2\big[\tfrac{1}{2}y_1\psi_1+\tfrac{3}{2}y_2\psi_2-\tfrac{1}{4}(y_1^2+3y_2^2)\psi_3\big],$$

$$m_3=a_2\mu^2\big[y_1y_2\psi_1+\tfrac{1}{2}(y_1^2-y_2^2)\psi_2+\tfrac{1}{6}(y_2^3-3y_1^2y_2)\psi_3\big],$$

$$m_4=a_2\big[(4h^2\mu(\lambda+2\mu)+2\mu^2y_2^2)\psi_2-\tfrac{2}{3}\mu^2y_2^3\psi_3\big],$$

$$m_5 = a_2\left[(4h^2\mu(\lambda+2\mu)+2\mu^2 y_1^2)\psi_1 - \tfrac{2}{3}\mu^2 y_1^3\psi_3\right],$$

$$m_6 = a_2\mu^2\left[\tfrac{1}{2}(y_2^2-y_1^2)\psi_1 + y_1 y_2\psi_2 + \tfrac{1}{6}(y_1^3-3y_1 y_2^2)\psi_3\right],$$

$$m_7 = a_2\left[(4h^2\mu(\lambda+2\mu)y_2 + \mu^2\tfrac{1}{6}(9y_1^2 y_2+y_2^3))\psi_1\right.$$
$$+\left(4h^2\mu(\lambda+2\mu)y_1 + \tfrac{1}{2}\mu^2(y_1^3+y_1 y_2^2)\right)\psi_2$$
$$\left.- \tfrac{1}{6}\mu^2(3y_1^3 y_2+y_1 y_2^3)\psi_3\right],$$

$$m_8 = a_2\left[(4h^2\mu(\lambda+2\mu)y_1 + \tfrac{1}{6}\mu^2(5y_1^3-3y_1 y_2^2))\psi_1\right.$$
$$- (4h^2\mu(\lambda+2\mu)y_2 + \tfrac{1}{2}\mu^2(y_1^2 y_2+y_2^3))\psi_2$$
$$\left.+ \tfrac{1}{24}\mu^2(-5y_1^4+6y_1^2 y_2^2+3y_2^4)\psi_3\right.\!,$$

$$m_9 = a_2\left[\left(4h^2\mu(\lambda+2\mu)y_2 + \tfrac{1}{2}\mu^2(y_1^2 y_2+y_2^3)\right)\psi_1\right.$$
$$+ (4h^2\mu(\lambda+2\mu)y_1 + \tfrac{1}{6}\mu^2(y_1^3+9y_1 y_2^2))\psi_2$$
$$\left.- \tfrac{1}{6}\mu^2(y_1^3 y_2+3y_1 y_2^3)\psi_3\right],$$

$$m_{10} = a_2\left[(4h^2\mu(\lambda+2\mu)y_1 + \tfrac{1}{2}\mu^2(y_1^3+y_1 y_2^2))\psi_1\right.$$
$$+ (-4h^2\mu(\lambda+2\mu)y_2 + \tfrac{1}{6}\mu^2(3y_1^2 y_2-5y_2^3))\psi_2$$
$$\left.- \tfrac{1}{24}\mu^2(3y_1^4+6y_1^2 y_2^2-5y_2^4)\psi_3\right],$$

$$m_{11} = a_2\left[\tfrac{1}{3}\mu^2(3y_1^2 y_2+y_2^3)\psi_1 + \tfrac{1}{3}\mu^2(y_1^3+3y_1 y_2^2)\psi_2\right.$$
$$\left.+ \left(4h^2\mu(\lambda+2\mu)y_1 y_2 - \tfrac{1}{3}\mu^2(y_1^3 y_2+y_1 y_2^3)\right)\psi_3\right],$$

$$m_{12} = a_2\left[\tfrac{2}{3}\mu^2(y_1^3\psi_1 - y_2^3\psi_2)\right.$$
$$\left.+ \left(2h^2\mu(\lambda+2\mu)(y_1^2-y_2^2) + \tfrac{1}{6}\mu^2(y_2^4-y_1^4)\right)\psi_3\right].$$

1.24 Remark. The above expressions of the residuals for $D(x_p,y)\psi(y)$ show that

$$\int_{\partial S}(D(x_p,y))^\infty\psi(y)\,ds(y) = 0$$

if and only if $\langle F,\psi\rangle = 0$; or, in view of (1.5), if and only if

$$\langle f^{(\alpha)},\psi\rangle = \int_{\partial S}(f^{(\alpha)})^\mathsf{T}\psi\,ds = \int_{\partial S}[\psi_\alpha(y)-y_\alpha\psi_3(y)]\,ds(y) = 0,$$

$$\langle f^{(3)},\psi\rangle = \int_{\partial S}(f^{(3)})^\mathsf{T}\psi\,ds = \int_{\partial S}\psi_3(y)\,ds(y) = 0.$$

These are, by implication, necessary and sufficient conditions to have

$$\int_{\partial S} D(x,y)\psi(y)\,ds(y) = \int_{\partial S} (D(x,y))^{\mathscr{A}}\,\psi(y)\,ds(y).$$

1.2.4 Components of Vector $P(x,y)\psi(y)$

We recall that $(P(x,y))^{\infty} = 0$, which means that $P(x,y) = (P(x,y))^{\mathscr{A}}$. The argument used in Sect. 1.2.3 applies here as well; that is, the asymptotic expansions of the entries of $P(x_p,y)\psi(y)$ are linear combinations of those of the columns $(P(x_p,y))^{(i)}$ of $P(x_p,y)$, with coefficients $\psi_i(y)$. In polar coordinates, these expansions are

$$\left(P(x_p,y)\psi(y)\right)_1 = \left((P(x_p,y))^{\mathscr{A}}\psi(y)\right)_1$$

$$= a_2\{r^{-1}\{2h^2\mu^3(n_2\psi_1 + n_1\psi_2)\sin\theta$$

$$+ 2h^2\mu^2[(2\lambda + 3\mu)n_1\psi_1 + (2\lambda + \mu)n_2\psi_2]\cos\theta$$

$$- 2h^2\mu^3(n_2\psi_1 + n_1\psi_2)\sin(3\theta)$$

$$+ 2h^2\mu^3(n_2\psi_2 - n_1\psi_1)\cos(3\theta)\}$$

$$+ r^{-2}\{[4h^2\mu^2(\mu n_2 y_1 + (\lambda + \mu)n_1 y_2)\psi_1$$

$$+ 4h^2\mu^2(\mu n_1 y_1 + (\lambda + \mu)n_2 y_2)\psi_2$$

$$+ 4h^2\mu^2(\lambda + 2\mu)n_2\psi_3]\sin(2\theta)$$

$$+ [4h^2\mu^2(\lambda + 2\mu)n_1 y_1\psi_1 + 4h^2\lambda\mu^2 n_2 y_1\psi_2$$

$$+ 4h^2\mu y^2(\lambda + 2\mu)n_1\psi_3]\cos(2\theta)$$

$$+ [-4h^2\mu^3(n_2 y_1 + n_1 y_2)\psi_1 + 4h^2\mu^3(n_2 y_2 - n_1 y_1)\psi_2]\sin(4\theta)$$

$$+ [4h^2\mu^3(n_2 y_2 - n_1 y_1)\psi_1 + 4h^2\mu^3(n_2 y_1 + n_1 y_2)\psi_2]\cos(4\theta)\}$$

$$+ r^{-3}\{[2h^2\mu^2(\mu n_2(3y_1^2 + y_2^2) + 8h^2(\lambda + 2\mu)n_2 + 2(2\lambda + 3\mu)n_1 y_1 y_2)\psi_1$$

$$+ 2h^2\mu^2(\mu n_1(3y_1^2 + y_2^2) + 2(2\lambda + \mu)n_2 y_1 y_2 + 8h^2(\lambda + 2\mu)n_1)\psi_2$$

$$+ 8h^2\mu^2(\lambda + 2\mu)(n_2 y_1 + n_1 y_2)\psi_3]\sin(3\theta)$$

$$+ 2h^2\mu^2(-2\mu n_2 y_1 y_2 + 8h^2(\lambda + 2\mu)n_1$$

$$+ (2\lambda + 5\mu)n_1 y_1^2 - (2\lambda + \mu)n_1 y_2^2)\psi_1$$

$$- 2h^2\mu^2(2\mu n_1 y_1 y_2 + 8h^2(\lambda + 2\mu)n_2$$

$$+ (\mu - 2\lambda)n_2 y_1^2 + (2\lambda + 3\mu)n_2 y_2^2)\psi_2$$
$$+ 8h^2\mu^2(\lambda + 2\mu)(n_1 y_1 - n_2 y_2)\psi_3]\cos(3\theta)$$
$$+ [6h^2\mu^3(-2n_1 y_1 y_2 + n_2(y_2^2 - y_1^2))\psi_1$$
$$+ 6h^2\mu^3(2n_2 y_1 y_2 + n_1(y_2^2 - y_1^2))\psi_2]\sin(5\theta)$$
$$+ [6h^2\mu^3(2n_2 y_1 y_2 + n_1(y_2^2 - y_1^2))\psi_1$$
$$+ 6h^2\mu^3(2n_1 y_1 y_2 + n_2(y_1^2 - y_2^2))\psi_2]\cos(5\theta)\} + O(r^{-4}),$$

$$\left(P(x_p, y)\,\psi(y)\right)_2 = \left((P(x_p, y))^{\mathscr{A}}\,\psi(y)\right)_2$$
$$= a_2\{r^{-1}\{[2h^2\mu^2(2\lambda + \mu)n_1\psi_1 + 2h^2\mu^2(2\lambda + 3\mu)n_2\psi_2]\sin\theta$$
$$+ 2h^2\mu^3(n_2\psi_1 + n_1\psi_2)\cos\theta$$
$$+ 2h^2\mu^3(-n_1\psi_1 + n_2\psi_2)\sin(3\theta)$$
$$+ 2h^2\mu^3(n_2\psi_2 + n_1\psi_2)\cos(3\theta)\}$$
$$+ r^{-2}\{[4h^2\mu^2(\mu n_2 y_2 + (\lambda + \mu)n_1 y_1)\psi_1$$
$$+ 4h^2\mu^2(\mu n_1 y_2 + (\lambda + \mu)n_2 y_1)\psi_2$$
$$+ 4h^2\mu^2(\lambda + 2\mu)n_1\psi_3]\sin(2\theta)$$
$$- [4h^2\lambda\mu^2 n_1 y_2\psi_1 + 4h^2\mu^2(\lambda + 2\mu)n_2 y_2\psi_2$$
$$+ 4h^2\mu^2(\lambda + 2\mu)n_2\psi_3]\cos(2\theta)$$
$$+ [4h^2\mu^3(n_2 y_2 - n_1 y_1)\psi_1 + 4h^2\mu^3(n_2 y_1 + n_1 y_2)\psi_2]\sin(4\theta)$$
$$+ [4h^2\mu^3(n_2 y_1 + n_1 y_2)\psi_1 + 4h^2\mu^3(n_1 y_1 - n_2 y_2)\psi_2]\cos(4\theta)\}$$
$$+ r^{-3}\{[2h^2\mu^2(2\mu n_2 y_1 y_2 + 8h^2(\lambda + 2\mu)n_1$$
$$+ (\mu - 2\lambda)n_1 y_2^2 + (2\lambda + 3\mu)n_1 y_1^2)\psi_1$$
$$+ 2h^2\mu^2(2\mu n_1 y_1 y_2 - 8h^2(\lambda + 2\mu)n_2$$
$$+ (2\lambda + \mu)n_2 y_1^2 - (2\lambda + 5\mu)n_2 y_2^2)\psi_2$$
$$+ 8h^2\mu^2(\lambda + 2\mu)(n_1 y_1 - n_2 y_2)\psi_3]\sin(3\theta)$$
$$+ [-2h^2\mu^2(\mu n_2(y_1^2 + 3y_2^2) + 2(2\lambda + \mu)n_1 y_1 y_2$$
$$+ 8h^2(\lambda + 2\mu)n_2)\psi_1$$
$$+ 2h^2\mu^2(\mu n_1(y_1^2 + 3y_2^2) + 8h^2(\lambda + 2\mu)n_1$$
$$+ 2(2\lambda + 3\mu)n_2 y_1 y_2)\psi_2$$
$$+ 8h^2\mu^2(\lambda + 2\mu)(n_2 y_1 + n_1 y_2)\psi_3]\cos(3\theta)$$
$$+ [6h^2\mu^3(2n_2 y_1 y_2 + n_1(y_2^2 - y_1^2))\psi_1$$

$$+ 6h^2\mu^3(2n_1y_1y_2 + n_2(y_1^2 - y_2^2))\psi_2]\sin(5\theta)$$
$$+ [6h^2\mu^3(2n_1y_1y_2 + n_2(y_1^2 - y_2^2))\psi_1$$
$$- 6h^2\mu^3(2n_2y_1y_2 + n_1(y_2^2 - y_1^2))\psi_2]\cos(5\theta)\}\} + O(r^{-4}),$$

$$\bigl(P(x_p,y)\psi(y)\bigr)_3 = \bigl((P(x_p,y))^{\mathscr{A}}\psi(y)\bigr)_3$$
$$= a_2\{-4h^2\mu^2(\lambda+\mu)(\ln r + 1)(n_1\psi_1 + n_2\psi_2)$$
$$- 2h^2\mu^3(n_2\psi_1 + n_1\psi_2)\sin(2\theta)$$
$$- 2h^2\mu^3(n_1\psi_1 - n_2\psi_2)\cos(2\theta)\}$$
$$+ r^{-1}\{[2h^2\mu^2(\mu n_2y_1 + (2\lambda+\mu)n_1y_2)\psi_1$$
$$+ 2h^2\mu^2(\mu n_1y_1 + (2\lambda+3\mu)n_2y_2)\psi_2$$
$$+ 4h^2\mu^2(\lambda+2\mu)n_2\psi_3]\sin\theta$$
$$+ [2h^2\mu^2(\mu n_2y_2 + (2\lambda+3\mu)n_1y_1)\psi_1$$
$$+ 2h^2\mu^2(\mu n_1y_2 + (2\lambda+\mu)n_2y_1)\psi_2$$
$$+ 4h^2\mu^2(\lambda+2\mu)n_1\psi_3]\cos\theta$$
$$+ [-2h^2\mu^3(n_2y_1 + n_1y_2)\psi_1$$
$$+ 2h^2\mu^3(-n_1y_1 + n_2y_2)\psi_2]\sin(3\theta)$$
$$+ [2h^2h^3(-n_1y_1 + n_2y_2)\psi_1$$
$$+ 2h^2\mu^3(n_2y_1 + n_1y_2)\psi_2]\cos(3\theta)\}$$
$$+ r^{-2}\{[2h^2\mu^2(\mu n_2(y_1^2 + y_2^2) + 2(\lambda+\mu)n_1y_1y_2)\psi_1$$
$$+ 2h^2\mu^2(\mu n_1(y_1 + y_2^2) + 2(\lambda+\mu)n_2y_1y_2)\psi_2$$
$$+ 4h^2\mu^2(\lambda+2\mu)(n_2y_1 + n_1y_2)\psi_3]\sin(2\theta)$$
$$+ 2h^2\mu^2(-\lambda n_1y_2^2 + (\lambda+2\mu)n_1y_1^2)\psi_1$$
$$- 2h^2\mu^2(-\lambda n_2y_1^2 + (\lambda+2\mu)n_2y_2^2)\psi_2$$
$$+ 4h^2\mu^2(\lambda+2\mu)(n_1y_1 - n_2y_2)\psi_3]\cos(2\theta)$$
$$+ [-2h^2\mu^3(2n_1y_1y_2 + n_2(y_1^2 - y_2^2))\psi_1$$
$$+ 2h^2\mu^2(2n_2y_1y_2 + n_1(y_2^2 - y_1^2))\psi_2]\sin(4\theta)$$
$$+ [2h^2\mu^3(2n_2y_1y_2 + n_1(y_2^2 - y_1^2))\psi_1$$
$$+ 2h^2\mu^3(2n_1y_1y_2 + n_2(y_1^2 - y_2^2))\psi_2]\cos(4\theta)\} + O(r^{-3}).$$

1.25 Remark. These expansions show that the class $\mathscr{A}$ pattern coefficients for $P(x_p,y)\psi(y) = \bigl(P(x_p,y)\bigr)^{\mathscr{A}}\psi(y)$ are

$$m_0 = 2a_2h^2\mu^3(n_2\psi_1 + n_1\psi_2),$$

$$m_1 = a_2 h^2 \mu^2 [(2\lambda + 3\mu)n_1 \psi_1 + (2\lambda + \mu)n_2 \psi_2],$$

$$m_2 = a_2 h^2 \mu^2 [(2\lambda + \mu)n_1 \psi_1 + (2\lambda + 3\mu)n_2 \psi_2],$$

$$m_3 = 2a_2 h^2 \mu^3 [(n_2 y_1 + n_1 y_2) \psi_1 + (n_1 y_1 - n_2 y_2) \psi_2],$$

$$m_4 = 4a_2 h^2 \mu^2 [\ln_1 y_2 \psi_1 + (\lambda + 2\mu)n_2 y_2 \psi_2 + (\lambda + 2\mu)n_2 \psi_3],$$

$$m_5 = 4a_2 h^2 \mu^2 [(\lambda + 2\mu)n_1 y_1 \psi_1 + \lambda n_2 y_1 \psi_2 + (\lambda + 2\mu)n_1 \psi_3],$$

$$m_6 = 2a_2 h^2 \mu^3 [(n2 y_2 - n_1 y_1) \psi_1 + (n_2 y_1 + n_1 y_2) \psi_2],$$

$$\begin{aligned}
m_7 = a_2 h^2 \mu^2 \{ & [8h^2(\lambda + 2\mu)n_2 + 2(2\lambda + 3\mu)n_1 y_1 y_2 + \mu n_2 (3y_1^2 + y_2^2)] \psi_1 \\
& + [8h^2(\lambda + 2\mu)n_1 + 2(2\lambda + \mu)n_2 y_1 y_2 + \mu n_1 (3y_1^2 + y_2^2)] \psi_2 \\
& + 4(\lambda + 2\mu)(n_2 y_1 + n_1 y_2) \psi_3 \},
\end{aligned}$$

$$\begin{aligned}
m_8 = a_2 h^2 \mu^2 \{ & [8h^2(\lambda + 2\mu)n_1 + (2\lambda + 5\mu)n_1 y_1^2 \\
& \quad - (2\lambda + \mu)n_1 y_2^2 - 2\mu n_2 y_1 y_2] \psi_1 \\
& + [-8h^2(\lambda + 2\mu)n_2 + (2\lambda - \mu)n_2 y_1^2 \\
& \quad - (2\lambda + 3\mu)n_2 y_2^2 - 2\mu n_1 y_1 y_2] \psi_2 \\
& + 4(\lambda + 2\mu)(n_1 y_1 - n_2 y_2) \psi_3 \},
\end{aligned}$$

$$\begin{aligned}
m_9 = a_2 h^2 \mu^2 \{ & [8h^2(\lambda + 2\mu)n_2 + 2(2\lambda + \mu)n_1 y_1 y_2 + \mu n_2 (y_1^2 + 3y_2^2)] \psi_1 \\
& + [8h^2(\lambda + 2\mu)n_1 + 2(2\lambda + 3\mu)n_2 y_1 y_2 + \mu n_1 (y_1^2 + 3y_2^2)] \psi_2 \\
& + 4(\lambda + 2\mu)(n_2 y_1 + n_1 y_2) \psi_3 \},
\end{aligned}$$

$$\begin{aligned}
m_{10} = a_2 h^2 \mu^2 \{ & [8h^2(\lambda + 2\mu)n_1 + 2(2\lambda + 3\mu)n_1 y_1^2 \\
& \quad + (\mu - 2\lambda)n_1 y_2^2 + 2\mu n_2 y_1 y_2] \psi_1 \\
& + [-8h^2(\lambda + 2\mu)n_2 + (2\lambda + \mu)n_2 y_1^2 \\
& \quad - (2\lambda + 5\mu)n_2 y_2^2 + 2\mu n_1 y_1 y_2] \psi_2 \\
& + 4(\lambda + 2\mu)(n_1 y_1 - n_2 y_2) \psi_3 \},
\end{aligned}$$

$$\begin{aligned}
m_{11} = 2a_2 h^2 \mu^2 \{ & [\mu n_2 (y_1^2 + y_2^2) + 2(\lambda + \mu)n_1 y_1 y_2] \psi_1 \\
& + [\mu n_1 (y_1^2 + y_2^2) + 2(\lambda + \mu)n_2 y_1 y_2] \psi_2 \\
& + 2(\lambda + 2\mu)(n_2 y_1 + n_1 y_2) \psi_3 \},
\end{aligned}$$

$$\begin{aligned}
m_{12} = 2a_2 h^2 \mu^2 \{ & [-\lambda n_1 y_2^2 + (\lambda + 2\mu)n_1 y_1^2] \psi_1 + [\lambda n_2 y_1^2 - (\lambda + 2\mu)n_2 y_2^2] \psi_2 \\
& + 2(\lambda + 2\mu)(n_1 y_1 - n_2 y_2) \psi_3 \}.
\end{aligned}$$

1.2.5 Integral Representation Formulas in S^-

The asymptotic expansions developed in Sects. 1.2.1–1.2.4 allow us to establish analogs of Theorems 1.7 and 1.10 in S^-.

1.26 Theorem. ([15], Theorem 3.13, p. 79) *If $u \in C^2(S^-) \cap C^1(\bar{S}^-) \cap \mathscr{A}^*$ is a solution of system* (1.3) *in S^-, then*

$$2 \int_{S^-} E(u,u) \, da = - \int_{\partial S} u^{\mathsf{T}} T u \, ds.$$

1.27 Theorem. ([15], Theorem 3.12, p. 79) *If $u \in C^2(S^-) \cap C^1(\bar{S}^-) \cap \mathscr{A}$ is a solution of system* (1.3) *in S^-, then*

$$-\int_{\partial S} \left[D(x,y) Tu(y) - P(x,y) u(y) \right] ds(y) = \begin{cases} 0, & x \in S^+, \\ \frac{1}{2} u(x), & x \in \partial S, \\ u(x), & x \in S^-. \end{cases} \tag{1.22}$$

1.3 Boundary Value Problems

Let $\mathscr{D}$, $\mathscr{N}$, and $\mathscr{R}$ be continuous 3×1 vector functions prescribed on ∂S, and let σ be a symmetric, positive definite, continuous 3×3 matrix function on ∂S. We consider the following interior and exterior Dirichlet, Neumann, and Robin boundary value problems [15, p. 80]:

(D^+) *Find $u \in C^2(S^+) \cap C^1(\bar{S}^+)$ satisfying*

$$Au(x) = 0, \quad x \in S^+,$$
$$u(x) = \mathscr{D}(x), \quad x \in \partial S.$$

(N^+) *Find $u \in C^2(S^+) \cap C^1(\bar{S}^+)$ satisfying*

$$Au(x) = 0, \quad x \in S^+,$$
$$Tu(x) = \mathscr{N}(x), \quad x \in \partial S.$$

(R^+) *Find $u \in C^2(S^+) \cap C^1(\bar{S}^+)$ satisfying*

$$Au(x) = 0, \quad x \in S^+,$$
$$Tu(x) + \sigma(x) u(x) = \mathscr{R}(x), \quad x \in \partial S.$$

(D^-) *Find $u \in C^2(S^-) \cap C^1(\bar{S}^-) \cap \mathscr{A}^*$ satisfying*

$$Au(x) = 0, \quad x \in S^-,$$
$$u(x) = \mathscr{D}(x), \quad x \in \partial S.$$

(N^-) *Find* $u \in C^2(S^-) \cap C^1(\bar{S}^-) \cap \mathscr{A}$ *satisfying*

$$Au(x) = 0, \quad x \in S^-,$$
$$Tu(x) = \mathscr{N}(x), \quad x \in \partial S.$$

(R^-) *Find* $u \in C^2(S^-) \cap C^1(\bar{S}^-) \cap \mathscr{A}^*$ *satisfying*

$$Au(x) = 0, \quad x \in S^-,$$
$$Tu(x) - \sigma(x)u(x) = \mathscr{R}(x), \quad x \in \partial S.$$

1.28 Definition. A solution as described above is called a *regular solution* of the corresponding problem.

1.29 Theorem. ([15], Theorem 6.5, p. 137; Theorem 6.6, p. 139; Theorem 6.7, p. 140; Theorem 6.9, p. 141) *Let* $\alpha \in (0,1)$.

(i) *The interior Dirichlet problem* (D^+) *has a unique regular solution for any prescribed* $\mathscr{D} \in C^{1,\alpha}(\partial S)$.

(ii) *The interior Neumann problem* (N^+) *is solvable for* $\mathscr{N} \in C^{0,\alpha}(\partial S)$ *if and only if*

$$\langle F, \mathscr{N} \rangle = 0; \tag{1.23}$$

explicitly, that is

$$\int_{\partial S} [\mathscr{N}_\beta(x) - x_\beta \mathscr{N}_3(x)] \, ds(x) = 0,$$
$$\int_{\partial S} \mathscr{N}_3(x) \, ds(x) = 0. \tag{1.24}$$

The regular solution is unique up to a 3×1 *vector of the form* (1.6).

(iii) *The interior Robin problem* (R^+) *has a unique regular solution for any prescribed* $\mathscr{R} \in C^{0,\alpha}(\partial S)$.

(iv) *The exterior Dirichlet problem* (D^-) *has a unique regular solution for any prescribed* $\mathscr{D} \in C^{1,\alpha}(\partial S)$.

(v) *The exterior Neumann problem* (N^-) *has a unique regular solution for a prescribed* $\mathscr{N} \in C^{0,\alpha}(\partial S)$ *if and only if condition* (1.23) *(or, what is the same,* (1.24)) *is satisfied.*

(vi) *The exterior Robin problem* (R^-) *has a unique regular solution for any prescribed* $\mathscr{R} \in C^{0,\alpha}(\partial S)$.

1.4 Layer Potentials

We introduce the *single-layer potential*

$$(V\psi)(x) = \int_{\partial S} D(x,y)\psi(y)\,ds(y), \quad x \in S^+ \cup S^-,$$

and the *double-layer potential*

$$(W\psi)(x) = \int_{\partial S} P(x,y)\psi(y)\,ds(y), \quad x \in S^+ \cup S^-,$$

where ψ is a density 3×1 vector function.

1.30 Remark. In view of our notational convention (see Definition 1.2), the potentials can also be expressed in the alternative form

$$\begin{aligned}
(V\psi)(x) &= \langle D(x,\cdot), \psi \rangle, \\
(W\psi)(x) &= \langle P(x,\cdot), \psi \rangle,
\end{aligned} \qquad x \in S^+ \cup S^-. \tag{1.25}$$

1.31 Theorem. ([15], Theorem 4.1, p. 83) *If $\psi \in C(\partial S)$, then $V\psi$ and $W\psi$ are analytic and satisfy system (1.3) in $S^+ \cup S^-$.*

1.32 Theorem. ([15], Theorem 4.2, p. 84) *If $\psi \in C(\partial S)$, then*

(i) *$W\psi \in \mathscr{A}$;*

(ii) *$V\psi \in \mathscr{A}$ if and only if $\langle F, \psi \rangle = 0$; that is, by (1.5), if and only if*

$$\int_{\partial S} [\psi_\alpha(x) - x_\alpha \psi_3(x)]\,ds(x) = 0,$$

$$\int_{\partial S} \psi_3(x)\,ds(x) = 0.$$

1.33 Theorem. ([15], Theorem 4.5, p. 85; Corollary 4.8, p. 89; Theorem 4.10, p. 90) *Let $\alpha \in (0,1)$.*

(i) *If $\psi \in C(\partial S)$, then $V\psi \in C^{0,\alpha}(\mathbb{R}^2)$.*

(ii) *If $\psi \in C^{0,\alpha}(\partial S)$, then*

$$\begin{aligned}
(W\psi)^+ &= -\tfrac{1}{2}\psi + W_0\psi \quad \text{on } \partial S, \\
(W\psi)^- &= \tfrac{1}{2}\psi + W_0\psi \quad \text{on } \partial S,
\end{aligned}$$

where the boundary operator W_0 is defined by

$$(W_0\psi)(x) = \int_{\partial S} P(x,y)\psi(y)\,ds(y), \quad x \in \partial S,$$

the integral being understood as principal value. With these limiting values on ∂S, $W\psi \in C^{1,\alpha}(\bar{S}^+)$ and $W\psi \in C^{1,\alpha}(\bar{S}^-)$.

(iii) *If $\psi \in C^{0,\alpha}(\partial S)$, then*

$$\big(T(V\psi)\big)^+ = \tfrac{1}{2}\psi + W_0^*\psi \quad \text{on } \partial S,$$

$$\big(T(V\psi)\big)^- = -\tfrac{1}{2}\psi + W_0^*\psi \quad \text{on } \partial S,$$

where the adjoint W_0^* of the operator W_0 is defined by

$$(W_0^*\psi)(x) = \int_{\partial S} T(\partial_x)D(x,y)\psi(y)\,ds(y), \quad x \in \partial S,$$

with the integral understood as principal value.

(iv) *If $\psi \in C^{1,\alpha}(\partial S)$, then $W\psi \in C^{1,\alpha}(\bar{S}^+)$ and $W\psi \in C^{1,\alpha}(\bar{S}^-)$, and*

$$T(W\psi)^+ = T(W\psi)^- \quad \text{on } \partial S.$$

1.34 Theorem. *The null space of the operators $\mathcal{V}$ and $\mathcal{W}$ defined on the set of solutions $u \in C^2(S^-) \cap C^1(\bar{S}^-) \cap \mathcal{A}^*$ of system (1.3) by*

$$(\mathcal{V}u)(x) = \int_{\partial S} D(x,y)(Tu)(y)\,ds(y) = \langle D(x,\cdot), Tu \rangle, \quad x \in S^-,$$

$$(\mathcal{W}u)(x) = \int_{\partial S} P(x,y)u(y)\,ds(y) = \langle P(x,\cdot), u \rangle, \quad x \in S^-,$$

is the vector space of rigid displacements in $\bar{S}^-$.

Proof. If $\mathcal{V}u = 0$ in S^-, then

$$V(Tu) = 0 \quad \text{in } S^-,$$

so

$$\big(TV(Tu)\big)^- = 0;$$

hence, by Theorem 1.33(iii),

$$\big(W_0^* - \tfrac{1}{2}I\big)(Tu) = 0 \quad \text{on } \partial S.$$

Theorem 3.22(i), p. 140 in [14] now yields $Tu = 0$; consequently, from Theorem 1.26 it follows that $E(u,u) = 0$ in S^-, so, according to Theorem 1.5, u is a rigid displacement in $\bar{S}^-$.

If $\mathcal{W}u = 0$ in S^-, then

$$(Wu)^- = 0;$$

that is, by Theorem 1.33(ii),

$$\left(W_0 + \tfrac{1}{2}I\right)u = 0 \quad \text{on } \partial S,$$

so, in view of Theorem 3.22(ii), p. 140 in [14], u is a rigid displacement in $\bar{S}^-$. $\square$

1.35 Remark. Technically speaking, the null space of $\mathscr{V}$ and $\mathscr{W}$ in Theorem 1.34 is $\mathscr{F}|_{\partial S}$, since these integral operators are constructed with the boundary values of u.

1.36 Remarks.

(i) Direct calculation shows that $T(Fc) = 0$. Conversely, by Theorems 1.7 (or 1.26) and 1.5, $Tu = 0$ implies that $u = Fc$. In other words,

$$Tu = 0 \quad \text{if and only if} \quad u = Fc.$$

(ii) A useful property of the double-layer potential W follows from (1.12) with $u = Fc$, and states that

$$\left(W(Fc)\right)(x) = \int\limits_{\partial S} P(x,y)(Fc)(y)\,ds(y) = \begin{cases} -(Fc)(x), & x \in S^+, \\ 0, & x \in S^-. \end{cases} \tag{1.26}$$

1.37 Remark. The procedure described in Remark 1.11 applies to a solution of system (1.3) in S^- as well, with one modification. If the third entry of the solution contains a $\ln r$ term, then the coefficient of this term (which supplies a class $\mathscr{A}$ vertical translation—see Remark 1.12) must be subtracted from the linear part of that entry before computation.

Chapter 2
Generalized Fourier Series

2.1 Introduction

There are many methods for computing approximate solutions of two-dimensional elliptic boundary value problems, each with its advantages and disadvantages. Thus, the finite difference and finite element techniques have the ability to generate the approximate solution directly over the entire domain, but the former is restricted to low-order differences and the latter to low-order finite-element approximations. Additionally, both suffer from the overhead imposed by calculations on a two-dimensional grid.

Traditional boundary integral approaches are significantly more efficient as they start by computing the solution on a one-dimensional curve. In these methods, however, the domain solution has to be constructed from its boundary approximation. These techniques are also based on integration that needs to cope with both weak and strong (Cauchy) singularities, which need special numerical handling.

The superiority of the generalized Fourier series method developed in this book resides in the avoidance of singular integrals, achieved by the choice of the collocation points off the domain boundary. This technique also dominates the competition through its convergence rate. Here, the numerical convergence of the underlying algorithm improves with the boundary smoothness, becoming exponential, therefore much more potent than any fixed finite-order method, when the contour is infinitely smooth. In what follows, we intend to justify our statement both analytically and numerically.

Let

$$\mathcal{G} = \{\varphi^{(i)}(x)\}_{i=1}^{\infty}$$

be a linearly independent set of functions which is complete in $L^2(\partial S)$, and let $\psi \in L^2(\partial S)$ be a function to be approximated computationally from the prescribed boundary data of a boundary value problem. We examine two types of approximation procedures.

© Springer Nature Switzerland AG 2020
C. Constanda, D. Doty, *The Generalized Fourier Series Method*, Developments in
Mathematics 65, https://doi.org/10.1007/978-3-030-55849-9_2

2.2 Orthonormalization Method

We consider three specific orthonormalization techniques: the classical Gram–Schmidt (CGS), the modified Gram–Schmidt (MGS), and the Householder reflections (HR). In all of them, $\mathscr{G}$ is orthonormalized as the set

$$G = \{\omega^{(i)}\}_{i=1}^{\infty},$$

which generates a traditional QR decomposition, written for each $n = 1, 2, \ldots$ as

$$\Phi = \Omega R, \tag{2.1}$$

with

$$\Phi = (\varphi^{(1)} \ \ldots \ \varphi^{(n)}), \quad \Omega = (\omega^{(1)} \ \ldots \ \omega^{(n)}),$$

and an upper-triangular nonsingular matrix $R = (r_{ij})$. Consequently,

$$\Omega = \Phi R^{-1},$$

where $R^{-1} = (\bar{r}_{ij})$ is also upper triangular. By components, this means that

$$\omega^{(i)} = \sum_{j=1}^{i} \bar{r}_{ij} \varphi^{(j)}, \quad i = 1, \ldots, n.$$

The set G is also linearly independent and complete in $L^2(\partial S)$, so

$$\psi = \sum_{i=1}^{\infty} c_i \omega^{(i)} = \sum_{i=1}^{\infty} \langle \psi, \omega^{(i)} \rangle \omega^{(i)}.$$

The approximation $\psi^{(n)}$ of ψ for $n = 1, 2, \ldots$ is now constructed in the form

$$
\begin{aligned}
\psi^{(n)} &= \sum_{i=1}^{n} \langle \psi, \omega^{(i)} \rangle \omega^{(i)} \\
&= \sum_{i=1}^{n} \langle \psi, \sum_{j=1}^{i} \bar{r}_{ij} \varphi^{(j)} \rangle \left(\sum_{h=1}^{i} \bar{r}_{ih} \varphi^{(h)} \right) \\
&= \sum_{i=1}^{n} \sum_{j=1}^{i} \bar{r}_{ij} \langle \psi, \varphi^{(j)} \rangle \left(\sum_{h=1}^{i} \bar{r}_{ih} \varphi^{(h)} \right),
\end{aligned}
\tag{2.2}
$$

with the numerical coefficients $\langle \psi, \varphi^{(j)} \rangle$, $j = 1, \ldots, n$, computed from the boundary condition of the problem.

2.1 Remark. Clearly, we have

$$\lim_{n \to \infty} \| \psi - \psi^{(n)} \| = 0. \tag{2.3}$$

2.2.1 Classical Gram–Schmidt Procedure (CGS)

In this method, each vector $\varphi^{(j)}$ is projected on the space orthogonal to the subspace

$$\text{span}\{\omega^{(1)}, \dots, \omega^{(j-1)}\}.$$

The projection operators are

$$P_1(\varphi^{(1)}) = I\varphi^{(1)},$$

$$P_{\omega_1 \dots \omega_{j-1}}(\varphi^{(j)}) = I(\varphi^{(j)}) - \sum_{i=1}^{j-1} \langle \omega^{(i)}, \varphi^{(j)} \rangle \omega^{(i)}, \quad j = 2, \dots, n.$$

Then

$$\omega^{(1)} = \frac{P_1 \varphi^{(1)}}{\|P_1 \varphi^{(1)}\|} = \frac{I\varphi^{(1)}}{\|I\varphi^{(1)}\|},$$

$$\omega^{(j)} = \frac{P_{\omega_1 \dots \omega_{j-1}} \varphi^{(j)}}{\|P_{\omega_1 \dots \omega_{j-1}} \varphi^{(j)}\|}, \quad j = 2, \dots, n.$$

The entries of the matrix R in (2.1) are

$$r_{11} = \|I\varphi^{(1)}\|,$$

$$r_{jj} = \|P_{\omega_1 \dots \omega_{j-1}} \varphi^{(j)}\|, \quad j = 2, \dots, n,$$

$$r_{ij} = \langle \omega^{(i)}, \varphi^{(j)} \rangle, \quad i = 1, \dots, j-1, \ j = 2, \dots, n,$$

$$r_{ij} = 0, \quad i = j+1, \dots, n, \ j = 1, \dots, n-1.$$

The obvious drawback when applying this method is that, as calculation progresses, the vector $\varphi^{(j)}$ tends to become increasingly 'parallel' to its projection on $\text{span}\{\omega^{(1)}, \dots, \omega^{(j-1)}\}$, so the difference between them becomes increasingly smaller in magnitude, which leads to computational ill-conditioning.

2.2.2 Modified Gram–Schmidt Procedure (MGS)

Due to the mutual orthogonality of the $\omega^{(i)}$, the projection operator of CGS can be re-factored in the form

$$P_1 = I,$$

$$P_{\omega_1 \dots \omega_{j-1}} = P_{\omega_{j-1}} \dots P_{\omega_1}, \quad j = 2, \dots, n.$$

The MGS is now constructed as

$$\omega^{(1)} = \frac{P_1 \varphi^{(1)}}{\|P_1 \varphi^{(1)}\|} = \frac{I\varphi^{(1)}}{I\varphi^{(1)}},$$

$$\omega^{(j)} = \frac{P_{\omega_1 \ldots \omega_{j-1}} \varphi^{(j)}}{\|P_{\omega_1 \ldots \omega_{j-1}} \varphi^{(j)}\|}$$

$$= \frac{P_{\omega_{j-1}} \ldots P_{\omega_1} \varphi^{(j)}}{\|P_{\omega_{j-1}} \ldots P_{\omega_1} \varphi^{(j)}\|}, \quad j = 2, \ldots, n.$$

MGS also generates a QR decomposition (2.1). Theoretically, R is the same for both CGS and MGS, since in both we have

$$\langle \omega^{(i)}, \varphi^{(j)} \rangle = r_{ij} = \langle \omega^{(i)}, P_{\omega_{j-1}} \ldots P_{\omega_1} \varphi^{(j)} \rangle. \tag{2.4}$$

However, computationally they are significantly different. Thus, the left-hand side in (2.4) is computed with the full magnitude of $\varphi^{(j)}$, whereas the right-hand side is computed with a vector function of a much smaller magnitude, which prevents the magnitude of the difference of the two in the inner product from becoming too small. This explains the computational superiority of MGS over CGS when floating-point arithmetic is used.

2.2.3 Householder Reflection Procedure (HR)

Both CGS and MGS suffer from an important computational shortcoming, in that the orthogonal projections $P_{\omega_1 \ldots \omega_{j-1}}$ are not invertible. By contrast, a Householder reflection H_j is invertible and orthogonal in the sense that $H_j^{-1} = H_j$, and is also norm preserving:

$$\|H_j \varphi\| = \|\varphi\|.$$

This makes HR the method of choice when performing orthogonal decompositions. The method was recently reinterpreted to allow its extension from finite-dimensional spaces to function spaces.

In HR, a target orthonormal set of vector functions $\{e^{(1)}, \ldots, e^{(n)}\}$ is selected, and the decomposition is defined as

$$(H_1 \varphi^{(1)} \quad H_2 H_1 \varphi^{(2)} \quad \ldots \quad H_n \ldots H_2 H_1 \varphi^{(n)}) = ER,$$

where

$$E = (e^{(1)} \quad \ldots \quad e^{(n)}).$$

What we need to know here is the structure of the H_j and formulas for the computation of matrix $R = (r_{ij})$.

An additional restriction, imported from the finite-dimensional HR, is imposed on this procedure. Specifically, we must have

$$H_j = I \quad \text{on } \operatorname{span}\{e^{(1)}, \ldots, e^{(j-1)}\}.$$

This leads to a QR decomposition (2.1) written in the form

$$\Theta = H_1 \ldots H_n \, ER = \Phi R. \tag{2.5}$$

Let $\bar{\varphi}_{j-1}^{(i)}$, $i = j, j+1, \ldots, n$, be the ith vector function constructed in step $j-1$. We start with $\bar{\varphi}_{j-1}^{(j)}$ and remove its components in $\operatorname{span}\{e^{(1)}, \ldots, e^{(j-1)}\}$, to produce $\bar{\bar{\varphi}}_{j-1}^{(j)}$. Next, we construct the vector function

$$\bar{\bar{\varphi}}_{j-1}^{(j)} - \|\bar{\bar{\varphi}}_{j-1}^{(j)}\| e^{(j)}.$$

Finally, we define the Householder reflection (through the hyperplane orthogonal to v_j) by

$$H_j(\varphi) = I\varphi - 2\frac{\langle v_j, \varphi \rangle}{\|v_j\|^2} v_j.$$

It can be verified that H_j satisfies all the necessary properties.

We now complete step j by applying H_j to the rest of the $\bar{\varphi}_{j-1}^{(i)}$, $i = j+1, \ldots, n$, generating

$$\bar{\varphi}_j^{(i)} = H_j(\bar{\varphi}_{j-1}^{(i)}).$$

This defines the entries of the upper-triangular matrix $R = (r_{ij})$ in (2.5) as

$$r_{jj} = \|\bar{\bar{\varphi}}_{j-1}^{(j)}\|, \quad j = 1, \ldots, n,$$
$$r_{ij} = \langle e^{(i)}, \bar{\varphi}_i^{(j)} \rangle, \quad i = 1, \ldots, j-1, \ j = 1, \ldots, n-1,$$
$$r_{ij} = 0, \quad i = j+1, \ldots, n, \ j = 1, \ldots, n-1.$$

2.2 Remark. CGS and MGS start with an incoming set $\{\varphi^{(1)}, \ldots, \varphi^{(n)}\}$ and predetermined rules for computing R, and construct an orthonormal set $\{\omega^{(1)}, \ldots, \omega^{(n)}\}$. HR starts with both an incoming set $\{\varphi^{(1)}, \ldots, \varphi^{(n)}\}$ and a preselected orthonormal set $\{e^{(1)}, \ldots, e^{(n)}\}$, and computes rules H_j and R that transform the former into the latter. As a consequence, numerical computation by HR can offer a dramatic improvement over the Gram–Schmidt procedures, subject to machine floating-point limitations.

2.3 Matrix Row Reduction Method (MRR)

In this technique, we expand the unknown function ψ in the original complete sequence $\mathcal{G}$:

$$\psi = \sum_{i=1}^{\infty} c_i \varphi^{(i)}, \quad c_i = \text{const},$$

so the approximation of ψ is

$$\psi^{(n)} = \sum_{i=1}^{n} c_i \varphi^{(i)}, \quad n = 1, 2, \ldots. \tag{2.6}$$

This yields

$$\langle \psi^{(n)}, \varphi^{(j)} \rangle = \sum_{i=1}^{n} c_i \langle \varphi^{(i)}, \varphi^{(j)} \rangle, \quad j = 1, 2, \ldots, n, \tag{2.7}$$

where the inner products $\langle \psi^{(n)}, \varphi^{(j)} \rangle = \beta_j$ are computed from the boundary condition of the problem.

2.3 Remark. Since $\mathscr{G}$ is complete, the convergence equality (2.3) holds here as well:

$$\lim_{n \to \infty} \| \psi - \psi^{(n)} \| = 0.$$

Introducing the $n \times n$ matrix $M = (M_{ij}) = (\langle \varphi^{(i)}, \varphi^{(j)} \rangle)$ and the column vectors $\beta = (\beta_j)$ and $c = (c_i)$, we can write (2.7) as

$$Mc = \beta,$$

which, since M is invertible, leads to

$$c = M^{-1} \beta. \tag{2.8}$$

The approximation $\psi^{(n)}$ is now given by (2.6), with the entries c_i of c obtained from system (2.8). Since $\mathscr{G}$ is a complete set in $L^2(\partial S)$, formula (2.3) holds here as well.

2.4 Remark. This method avoids the computational issues arising from the orthonormalization of the sequence $\mathscr{G}$.

2.5 Remark. The computational algorithm for a specific boundary value problem may require additional restrictions, caused by the nature of the problem, to be imposed on the approximate solution. Such refinements are addressed in the chapters where they occur.

Chapter 3
Interior Dirichlet Problem

3.1 Computational Algorithm

The interior Dirichlet problem consists of the equation and boundary condition

$$Au(x) = 0, \quad x \in S^+,$$
$$u(x) = \mathscr{D}(x), \quad x \in \partial S, \tag{3.1}$$

where the vector function $\mathscr{D} \in C^{1,\alpha}(\partial S)$, $\alpha \in (0,1)$, is prescribed.

The representation formulas (1.12) in this case become

$$u(x) = \int_{\partial S} D(x,y)Tu(y)\,ds(y) - \int_{\partial S} P(x,y)\mathscr{D}(y)\,ds(y), \quad x \in S^+, \tag{3.2}$$

$$0 = \int_{\partial S} D(x,y)Tu(y)\,ds(y) - \int_{\partial S} P(x,y)\mathscr{D}(y)\,ds(y), \quad x \in S^-. \tag{3.3}$$

The proof of the next assertion can be found in any elementary text on functional analysis. (See, for example, [25], Theorem 3.6-2, p. 169.)

3.1 Theorem. *If X is a Hilbert space, then $\mathscr{X} \subset X$ is a complete set in X if and only if the orthogonal complement $\mathscr{X}^\perp$ of $\mathscr{X}$ in X consists of the zero vector alone.*

3.2 Remark. Let $M(x,y)$ be either $D(x,y)$ or $P(x,y)$. For all problems in S^+ we have the following obvious notational equivalences:

$$M^{(j)}(x,y) = \left(M(x,y)\right)^{(j)};$$
$$M_{(i)}(x,y) = \left(M(x,y)\right)_{(i)};$$
$$M_{ij}(x,y) = \left(M(x,y)\right)_{ij} = \left(M^{(j)}(x,y)\right)_i = \left(\left(M(x,y)\right)^{(j)}\right)_i$$
$$= \left(M_{(i)}(x,y)\right)_j = \left(\left(M(x,y)\right)_{(i)}\right)_j;$$

© Springer Nature Switzerland AG 2020
C. Constanda, D. Doty, *The Generalized Fourier Series Method*, Developments in
Mathematics 65, https://doi.org/10.1007/978-3-030-55849-9_3

$$M^{\mathsf{T}}(x,y) = \big(M(x,y)\big)^{\mathsf{T}};$$

$$\big(M^{\mathsf{T}}(x,y)\big)^{(j)} = (M^{\mathsf{T}})^{(j)}(x,y) = \big(M_{(j)}(x,y)\big)^{\mathsf{T}} = (M_{(j)})^{\mathsf{T}}(x,y);$$

$$\big(M^{\mathsf{T}}(x,y)\big)_{(j)} = (M^{\mathsf{T}})_{(j)}(x,y) = \big(M^{(j)}(x,y)\big)^{\mathsf{T}} = (M^{(j)})^{\mathsf{T}}(x,y);$$

$$(M^{\mathsf{T}})_{ij}(x,y) = \big((M(x,y))^{\mathsf{T}}\big)_{ij} = \big(M(x,y)\big)_{ji} = M_{ji}(x,y).$$

This notation remains valid when x and y are replaced by other pairs of distinct points.

Let ∂S_* be a simple, closed, C^2-curve surrounding $\bar{S}^+$, let $\{x^{(k)}\}_{k=1}^{\infty}$ be a set of points densely distributed on ∂S_*, and consider the sequence of vector functions on ∂S defined by

$$\mathscr{G} = \{f^{(i)}, \ i = 1,2,3\} \cup \{\varphi^{(jk)}, \ j = 1,2,3, \ k = 1,2,\ldots\},$$

where

$$\varphi^{(jk)}(x) = D^{(j)}(x,x^{(k)}) = (D_{(j)})^{\mathsf{T}}(x^{(k)},x), \quad j = 1,2,3, \ k = 1,2,\ldots, \tag{3.4}$$

with the second expression derived from the symmetry (1.9).

3.3 Theorem. *The set $\mathscr{G}$ is linearly independent on ∂S and complete in $L^2(\partial S)$.*

Proof. Assuming the opposite, suppose that there are a positive integer N and real numbers c_i and c_{jk}, $i,j = 1,2,3$, $k = 1,2,\ldots,N$, not all zero, such that

$$\sum_{i=1}^{3} c_i f^{(i)}(x) + \sum_{j=1}^{3} \sum_{k=1}^{N} c_{jk} \varphi^{(jk)}(x) = 0, \quad x \in \partial S. \tag{3.5}$$

We set

$$g(x) = \sum_{i=1}^{3} c_i f^{(i)}(x) + \sum_{j=1}^{3} \sum_{k=1}^{N} c_{jk} D^{(j)}(x,x^{(k)}), \quad x \in S^+, \tag{3.6}$$

and see that, by (3.4), (3.5), and Theorem 1.9,

$$Ag = 0 \quad \text{in } S^+,$$
$$g = 0 \quad \text{on } \partial S;$$

that is, g is the (unique) regular solution of the homogeneous Dirichlet problem in S^+. According to Theorem 1.29(i), $g = 0$ in $\bar{S}^+$, so, by analyticity,

$$g = 0 \quad \text{in } \bar{S}_*^+. \tag{3.7}$$

Let $x^{(p)}$ be any one of the points $x^{(1)}, \ldots, x^{(N)}$. By (3.6) and (3.4), for $x \in \bar{S}_*^+$,

$$g_h(x) = \sum_{i=1}^{3} c_i f_h^{(i)}(x) + \sum_{j=1}^{3} \sum_{k=1}^{N} c_{jk} D_{hj}(x,x^{(k)}), \quad h = 1,2,3,$$

and we see that, in view of (1.11), as $x \to x^{(p)}$, all the terms on the right-hand side remain bounded except $c_{hp}D_{hh}(x, x^{(p)})$, which is $O(\ln|x - x^{(p)}|)$. This contradicts (3.7), therefore, repeating the argument for every $x^{(k)}$, $k = 1, 2, \ldots, N$, we conclude that all the coefficients c_{jk} in (3.6) must be zero. The linear independence of the $f^{(i)}$, $i = 1, 2, 3$, now implies that the c_i are also zero. This means that the set $\mathscr{G}$ is linearly independent on ∂S.

For the second part of the proof, suppose that there is $q \in L^2(\partial S)$ such that

$$\langle q, f^{(i)} \rangle = \int_{\partial S} (f^{(i)})^{\mathsf{T}} q \, ds = 0, \quad i = 1, 2, 3, \tag{3.8}$$

$$\langle q, \varphi^{(jk)} \rangle = \int_{\partial S} (\varphi^{(jk)})^{\mathsf{T}} q \, ds = 0, \quad j = 1, 2, 3, \ k = 1, 2, \ldots. \tag{3.9}$$

According to (3.4), equality (3.9) is the same as

$$\int_{\partial S} D(x^{(k)}, y) q(y) \, ds(y) = 0, \quad k = 1, 2, \ldots. \tag{3.10}$$

Next, by Theorem 4.18(i), p.99 in [15], the single-layer potential with L^2-density

$$(Vq)(x) = \int_{\partial S} D(x, y) q(y) \, ds(y), \quad x \in S^+ \cup S^-,$$

is continuous on ∂S_*, so, since the points $x^{(k)}$, $k = 1, 2, \ldots$, are densely distributed on ∂S_*, from (3.10) we deduce that $Vq = 0$ on ∂S_*. By (3.8) and Theorem 4.18(ii),(iii), p. 99 in [15],

$$A(Vq) = 0 \quad \text{in } S_*^-,$$
$$Vq = 0 \quad \text{on } \partial S_*,$$
$$Vq \in \mathscr{A},$$

which shows that Vq is the (unique) solution of the homogeneous Dirichlet problem in $\bar{S}_*^-$; hence, by Theorem 1.29(iv),

$$Vq = 0 \quad \text{in } \bar{S}_*^-.$$

Since the single-layer potential Vq is analytic in $S^+ \cup S^-$, we have

$$Vq = 0 \quad \text{in } \bar{S}^-.$$

Consequently, $(T(Vq))^- = 0$, so from Theorem 4.20, p. 99 in [15] it follows that

$$-\frac{1}{2} q(x) + \int_{\partial S} T(\partial_x) D(x, y) q(y) \, ds(y) = 0 \quad \text{for a.a. } x \in \partial S, \tag{3.11}$$

with the integral above understood as principal value. According to Theorem 6.12, p. 145 in [15], $q \in C^{0,\alpha}(\partial S)$ for any $\alpha \in (0,1)$, so the integral equation (3.11) is also satisfied pointwise on ∂S. By Theorem 3.22, p. 140 in [14], the null space of the operator defined by this equation consists of the zero vector alone, so $q = 0$.

Given that $L^2(\partial S)$ is a Hilbert space, Theorem 3.1 now implies that the set $\mathscr{G}$ is complete in $L^2(\partial S)$. □

We order the elements of $\mathscr{G}$ as the sequence

$$f^{(1)}, f^{(2)}, f^{(3)}, \varphi^{(11)}, \varphi^{(21)}, \varphi^{(31)}, \varphi^{(12)}, \varphi^{(22)}, \varphi^{(32)}, \ldots$$

and re-index them, noticing that for each $i = 4, 5, \ldots$, there is a unique pair $\{j, k\}$, $j = 1, 2, 3$, $k = 1, 2, \ldots$, such that $i = j + 3k$. Then

$$\mathscr{G} = \{\varphi^{(1)}, \varphi^{(2)}, \varphi^{(3)}, \varphi^{(4)}, \varphi^{(5)}, \varphi^{(6)}, \varphi^{(7)}, \varphi^{(8)}, \varphi^{(9)}, \ldots\}, \qquad (3.12)$$

where

$$\varphi^{(i)} = f^{(i)}, \quad i = 1, 2, 3, \qquad (3.13)$$

and for $i = 4, 5, \ldots$,

$$\varphi^{(i)} = \varphi^{(jk)}, \quad i = j + 3k, \ j = 1, 2, 3, \ k = 1, 2, \ldots. \qquad (3.14)$$

3.1.1 Row Reduction Method

Our intention is to construct a nonsingular linear system of finitely many algebraic equations to approximate the solution u of (3.1).

Rewriting the representation formulas (3.2) and (3.3) as

$$u(x) = \langle D(x, \cdot), Tu \rangle - \langle P(x, \cdot), \mathscr{D} \rangle, \quad x \in S^+, \qquad (3.15)$$
$$\langle D(x, \cdot), Tu \rangle = \langle P(x, \cdot), \mathscr{D} \rangle, \quad x \in S^-, \qquad (3.16)$$

we aim to use (3.16) to approximate Tu (on ∂S), and then to approximate u in S^+ by means of (3.15).

First, we note that, by Theorem 1.34, the null spaces of the two operators defined by the left-hand side and the right-hand side in (3.16) coincide with $\mathscr{F}|_{\partial S}$; that is, we have

$$\langle D(x, \cdot), Tf^{(i)} \rangle = \langle P(x, \cdot), f^{(i)} \rangle = 0, \quad x \in S^-, \quad i = 1, 2, 3.$$

To guarantee a unique solution for the system that we intend to design, we need to replace the operators on both sides of (3.16) by new ones with a zero null space; in other words, we need to eliminate the rigid displacements $f^{(i)}$ from the formulation of that equality. The obvious way to do this is to seek the solution u of (3.1) in the form

$$u = u^{(c)} + Fa \quad \text{in } S^+,$$

where

$$Fa = \sum_{j=1}^{3} a_j f^{(j)}$$

and the 'clean' component $u^{(c)}$ is such that $u^{(c)}|_{\partial S}$ is orthogonal to the space of the rigid displacements on the boundary; that is,

$$\langle u^{(c)}, f^{(i)} \rangle = 0, \quad i = 1, 2, 3. \tag{3.17}$$

Then the boundary data function also splits as

$$\mathcal{D} = \mathcal{D}^{(c)} + Fa, \quad \mathcal{D}^{(c)} = u^{(c)}|_{\partial S}. \tag{3.18}$$

From (3.17) and (3.18) it follows that

$$\langle \mathcal{D}^{(c)}, f^{(i)} \rangle = \langle u^{(c)}, f^{(i)} \rangle = 0, \quad i = 1, 2, 3,$$

so, by (3.18),

$$\langle \mathcal{D}, f^{(i)} \rangle = \langle \mathcal{D}^{(c)}, f^{(i)} \rangle + \langle Fa, f^{(i)} \rangle$$
$$= \langle Fa, f^{(i)} \rangle = \sum_{j=1}^{3} a_j \langle f^{(j)}, f^{(i)} \rangle, \quad i = 1, 2, 3,$$

or, explicitly,

$$\begin{pmatrix} \langle f^{(1)}, f^{(1)} \rangle & \langle f^{(1)}, f^{(2)} \rangle & \langle f^{(1)}, f^{(3)} \rangle \\ \langle f^{(2)}, f^{(1)} \rangle & \langle f^{(2)}, f^{(2)} \rangle & \langle f^{(2)}, f^{(3)} \rangle \\ \langle f^{(3)}, f^{(1)} \rangle & \langle f^{(3)}, f^{(2)} \rangle & \langle f^{(3)}, f^{(3)} \rangle \end{pmatrix} \begin{pmatrix} a_1 \\ a_2 \\ a_3 \end{pmatrix} = \begin{pmatrix} \langle \mathcal{D}, f^{(1)} \rangle \\ \langle \mathcal{D}, f^{(2)} \rangle \\ \langle \mathcal{D}, f^{(3)} \rangle \end{pmatrix}.$$

This is a 3×3 system with a unique solution a, which determines the rigid displacement Fa when the exact solution u is not known. As it turns out, in this case knowing Fa is not necessary for computing an approximation of u in S^+.

Now

$$Tu = Tu^{(c)} + T(Fa) = Tu^{(c)} \quad \text{on } \partial S, \tag{3.19}$$

so

$$\langle D(x, \cdot), Tu \rangle = \langle D(x, \cdot), Tu^{(c)} \rangle, \quad x \in S^+ \cup S^-. \tag{3.20}$$

In view of (3.19), we set

$$\psi = Tu = Tu^{(c)} \quad \text{on } \partial S. \tag{3.21}$$

Since the $x^{(k)}$ are points in S^-, from (3.3), (3.20), and (3.21) it follows that

$$\int_{\partial S} D(x^{(k)}, x) \psi(x) \, ds(x) = \int_{\partial S} P(x^{(k)}, x) \mathcal{D}(x) \, ds(x),$$

which, by (3.4) and (3.14), leads to

$$\langle \varphi^{(i)}, \psi \rangle = \langle \varphi^{(jk)}, \psi \rangle = \int_{\partial S} D_{(j)}(x^{(k)}, x) \psi(x)\, ds(x) = \int_{\partial S} P_{(j)}(x^{(k)}, x) \mathscr{D}(x)\, ds(x),$$

$$j = 1, 2, 3, \quad k = 1, 2, \ldots, \quad i = 4, 5, \ldots,$$

or, what is the same,

$$\langle \varphi^{(i)}, \psi \rangle = \langle (D_{(j)})^{\mathsf{T}}(x^{(k)}, \cdot), \psi \rangle = \langle (P_{(j)})^{\mathsf{T}}(x^{(k)}, \cdot), \mathscr{D} \rangle. \tag{3.22}$$

Also, since, according to (3.21), ψ is the Neumann boundary data of the solution of (3.1), we have

$$\langle \psi, \varphi^{(i)} \rangle = \langle \psi, f^{(i)} \rangle = 0, \quad i = 1, 2, 3. \tag{3.23}$$

In view of Theorem 3.3, there is a unique expansion

$$\psi = \sum_{h=1}^{\infty} c_h \varphi^{(h)},$$

so we can seek an approximation of ψ of the form

$$\psi^{(n)} = (Tu)^{(n)} = (Tu^{(c)})^{(n)} = \sum_{h=1}^{n} c_h \varphi^{(h)}. \tag{3.24}$$

3.4 Remark. For symmetry and ease of computation, in the construction of the approximation $\psi^{(n)}$ we use the subsequence with $n = 3N + 3$, where N is the number of points $x^{(k)}$ selected on ∂S_*. This choice feeds into the computation, for each $k = 1, 2, \ldots, N$, all three columns/rows of D involved in the definition (3.4) of the vector functions $\varphi^{(i)}$, $i = 4, 5, \ldots, 3N + 3$, with the re-indexing (3.14).

Since ψ satisfies (3.23), it is reasonable to ask its approximation $\psi^{(n)}$ to satisfy the same equality; that is,

$$\langle \psi^{(n)}, f^{(i)} \rangle = 0, \quad i = 1, 2, 3. \tag{3.25}$$

Combining (3.24) and (3.25), and then (3.24) and (3.22), we obtain the approximate equalities

$$\sum_{h=1}^{n} c_h \langle f^{(i)}, \varphi^{(h)} \rangle = 0, \quad i = 1, 2, 3, \tag{3.26}$$

$$\sum_{h=1}^{n} c_h \langle \varphi^{(i)}, \varphi^{(h)} \rangle = \langle (P_{(j)})^{\mathsf{T}}(x^{(k)}, \cdot), \mathscr{D} \rangle,$$

$$j = 1, 2, 3, \quad k = 1, 2, \ldots, N, \quad i = 4, 5, \ldots, n = 3N + 3. \tag{3.27}$$

Equations (3.26) and (3.27) form our system for determining the $3N + 3$ coefficients c_h. In expanded form, this system is

$$
\begin{pmatrix}
\langle f^{(1)}, \varphi^{(1)} \rangle & \langle f^{(1)}, \varphi^{(2)} \rangle & \cdots & \langle f^{(1)}, \varphi^{(n)} \rangle \\
\langle f^{(2)}, \varphi^{(1)} \rangle & \langle f^{(2)}, \varphi^{(2)} \rangle & \cdots & \langle f^{(2)}, \varphi^{(n)} \rangle \\
\langle f^{(3)}, \varphi^{(1)} \rangle & \langle f^{(3)}, \varphi^{(2)} \rangle & \cdots & \langle f^{(3)}, \varphi^{(n)} \rangle \\
\langle \varphi^{(4)}, \varphi^{(1)} \rangle & \langle \varphi^{(4)}, \varphi^{(2)} \rangle & \cdots & \langle \varphi^{(4)}, \varphi^{(n)} \rangle \\
\vdots & \vdots & & \vdots \\
\langle \varphi^{(n)}, \varphi^{(1)} \rangle & \langle \varphi^{(n)}, \varphi^{(2)} \rangle & \cdots & \langle \varphi^{(n)}, \varphi^{(n)} \rangle
\end{pmatrix}
\begin{pmatrix}
c_1 \\ c_2 \\ c_3 \\ c_4 \\ \vdots \\ c_n
\end{pmatrix}
$$

$$
=
\begin{pmatrix}
0 \\
0 \\
0 \\
\langle (P_{(1)})^{\mathsf{T}}(x^{(1)}, \cdot), \mathscr{D} \rangle \\
\vdots \\
\langle (P_{(3)})^{\mathsf{T}}(x^{(N)}, \cdot), \mathscr{D} \rangle
\end{pmatrix}. \tag{3.28}
$$

It is easily checked that system (3.28) is nonsingular, therefore, it has a unique solution that generates the approximation (3.24) of ψ.

3.5 Remark. We have deliberately written $f^{(i)}$ instead of $\varphi^{(i)}$ in the inner products in the first three rows in (3.28), to emphasize that those equations have a different source than the rest.

By (3.21), representation formula (3.15) can be written as

$$
u(x) = \langle D(x, \cdot), \psi \rangle - \langle P(x, \cdot), \mathscr{D} \rangle, \quad x \in S^+,
$$

so we construct the vector function

$$
u^{(n)}(x) = \langle D(x, \cdot), \psi^{(n)} \rangle - \langle P(x, \cdot), \mathscr{D} \rangle, \quad x \in S^+. \tag{3.29}
$$

3.6 Theorem. *The vector function $u^{(n)}$ defined by (3.29) is an approximation of the solution u of problem (3.1) in the sense that $u^{(n)} \to u$ uniformly on any closed subdomain S' of S^+.*

Proof. From (3.2) and (3.29) we find that

$$
\left| u(x) - u^{(n)}(x) \right| \le \sum_{i=1}^{3} \left| u_i(x) - u_i^{(n)}(x) \right|
$$

$$
\le \sum_{i=1}^{3} \int_{\partial S} \left| D_{(i)}(x, y) \left[\psi(y) - \psi^{(n)}(y) \right] \right| ds(y)
$$

$$
\le \sum_{i=1}^{3} \| D_{(i)}(x, \cdot) \| \, \| \psi - \psi^{(n)} \|, \quad x \in S^+.
$$

Since, as seen from formula (3.30), p. 75 in [15], the $\| D_{(i)}(x, \cdot) \|$ are uniformly bounded on any closed subdomain $S' \subset S^+$ and, by (2.3), $\| \psi - \psi^{(n)} \| \to 0$ as $n \to \infty$, we conclude that $u^{(n)} \to u$, uniformly on S'. $\qquad \square$

3.1.2 Orthonormalization Method

After the re-indexing (3.13), (3.14), we orthonormalize the set $\{\varphi^{(i)}\}_{i=1}^{n}$ as

$$\{\omega^{(i)}\}_{i=1}^{n}, \quad n = 3N + 3.$$

This yields the *QR* factorization

$$\Phi = \Omega R, \tag{3.30}$$

where the matrices Φ, Ω, and the nonsingular, upper-triangular $n \times n$ matrix R are described in Sect. 2.2.

We rewrite system (3.28) in the form

$$Mc = \beta,$$

where

$$M = (M_{ij}) = (\langle \varphi^{(i)}, \varphi^{(j)} \rangle), \quad i, j = 1, 2, \ldots, n,$$

$c = (c_1, c_2, \ldots, c_n)^{\mathsf{T}}$, and β is the column vector of the numbers on the right-hand side of (3.28). Then

$$c = M^{-1}\beta,$$

and the approximation $\psi^{(n)}$ of $Tu^{(c)}$ is given by

$$\psi^{(n)} = \Phi c = \Phi M^{-1}\beta = \Omega R M^{-1}\beta.$$

Since the direct computation of M^{-1} is very inefficient, we prefer to avoid it by using the orthonormality of the set Ω. According to our notation,

$$M = (\langle \varphi^{(i)}, \varphi^{(j)} \rangle) = \langle \Phi^{\mathsf{T}}, \Phi \rangle,$$

so, by (3.30),

$$M = \langle R^{\mathsf{T}}\Omega^{\mathsf{T}}, \Omega R \rangle = R^{\mathsf{T}}\langle \Omega^{\mathsf{T}}, \Omega \rangle R = R^{\mathsf{T}}IR = R^{\mathsf{T}}R.$$

Hence,

$$RM^{-1} = R(R^{\mathsf{T}}R)^{-1} = RR^{-1}(R^{\mathsf{T}})^{-1} = (R^{\mathsf{T}})^{-1},$$

which leads to

$$\psi^{(n)} = \Omega(R^{\mathsf{T}})^{-1}\beta. \tag{3.31}$$

R is upper triangular, so $(R^{\mathsf{T}})^{-1} = (\bar{r}_{ih})$ is lower triangular and can be computed by forward substitution. This reduces (3.31) to

$$\psi^{(n)} = \sum_{i=1}^{n} \left(\sum_{h=1}^{i} \bar{r}_{ih}\beta_h \right) \omega^{(i)}, \quad n = 3N + 3,$$

which, given that $\beta_1 = \beta_2 = \beta_3 = 0$, simplifies further to

$$\psi^{(n)} = \sum_{i=4}^{n} \sum_{h=4}^{i} \bar{F}_{ih} \beta_h \omega^{(i)}$$

$$= \sum_{i=4}^{n} \sum_{h=4}^{i} \bar{F}_{ih} \langle (P_{(j)})^{\mathsf{T}} (x^{(k)}, \cdot), \mathscr{D} \rangle \omega^{(i)}, \quad n = 3N + 3.$$

Once $\psi^{(n)}$ has been computed, we can use (3.29) to construct the approximation $u^{(n)}$ of the solution u of (3.1).

3.7 Remark. The orthonormalization method has the advantage that it computes directly each of the elements used in the theoretical construction. The matrix row reduction method is easier to set up, but does not illustrate the theoretical procedure.

3.2 Numerical Example 1

Let S^+ be the disk of radius 1 centered at origin, and suppose that, after rescaling and non-dimensionalization, the physical and geometric parameters of the plate are

$$h = 0.5, \quad \lambda = \mu = 1.$$

We choose the boundary condition, written in polar coordinates, in (3.1) to be

$$\mathscr{D}(x_p) = \begin{pmatrix} 3 + 8\cos\theta + 11\cos(2\theta) - 2\cos(3\theta) \\ 4\sin\theta - 7\sin(2\theta) - 2\sin(3\theta) \\ 6 + 11\cos\theta - 20\cos(2\theta) - 3\cos(3\theta) \end{pmatrix}.$$

It is easily verified that the exact solution of the problem for this function is

$$u(x) = \begin{pmatrix} -1 + 14x_1 + 15x_1^2 - 7x_2^2 - 8x_1^3 \\ -2x_2 - 14x_1x_2 + 8x_2^3 \\ 9 + 13x_1 - 25x_1^2 + 19x_2^2 - 5x_1^3 + 7x_1x_2^2 + 2x_1^4 - 2x_2^4 \end{pmatrix}.$$

3.8 Remark. Applying the scheme described in Remark 1.12 and writing

$$w = (\alpha_1 x_1 + \alpha_2 x_2 + \alpha_3, \, \beta_1 x_1 + \beta_2 x_2 + \beta_3, \, \gamma_1 x_1 + \gamma_2 x_2 + \gamma_3)^{\mathsf{T}},$$

we have, successively,

$$v = (-1 + 14x_1, -2x_2, 9 + 13x_1)^{\mathsf{T}},$$
$$Av = (12 - 14x_1, 2x_2, 12)^{\mathsf{T}},$$
$$Aw = (-\alpha_1 x_1 - \alpha_2 x_2 - \alpha_3 - \gamma_1, \, -\beta_1 x_1 - \beta_2 x_2 - \beta_3 - \gamma_2, \, \alpha_1 + \beta_2)^{\mathsf{T}},$$

and the equality $Av = Aw$ yields

$$w = (12 + 14x_1, -2x_2, 0)^\mathsf{T} - \gamma_1 f^{(1)} - \gamma_2 f^{(2)} + \gamma_3 f^{(3)}.$$

We now set $\gamma_1 = \gamma_2 = \gamma_3 = 0$ and find that the rigid displacement component of the solution in this example is

$$f = -13f^{(1)} + 9f^{(3)}.$$

We select the auxiliary curve ∂S_* to be the circle of radius 2 centered at the origin.

3.9 Remark. This choice is based on the fact that

(i) if ∂S_* is too far away from ∂S, then the sequence $\mathscr{G}$ becomes 'less linearly independent';

(ii) if ∂S_* is too close to ∂S, then $\mathscr{G}$ becomes increasingly sensitive to the singularities of D and P on the boundary.

3.10 Remark. Clearly, the accuracy of the approximation depends on the selection of the set of points $x^{(k)}$ on ∂S_*. We make the reasonable choice of spacing these points uniformly around ∂S_*; that is, for $n = 1, 2, \ldots$,

$$\{x^{(k)} : k = 1, 2, \ldots, n\}_{\text{Cartesian}} = \left\{ \left(2, \frac{2\pi k}{n} \right) : k = 1, 2, \ldots, n \right\}_{\text{Polar}}.$$

3.11 Remark. Obviously, $\{x^{(k)}\}_{k=1}^{\infty}$ is the set of all points on ∂S_* whose polar angle is of the form $2\pi\alpha$, where α is any rational number such that $0 < \alpha \leq 1$.

3.12 Remark. We have performed floating-point computation with machine precision of approximately 16 digits. The most sensitive part of the process is the evaluation of integrals in the inner products, for which we set a target of 11 significant digits.

We approximate the solution of problem (3.1) in this example by means of the MGS method.

3.2.1 Graphical Illustrations

The graphs of the components of $u^{(51)}$ computed from $\psi^{(51)}$ (that is, with 16 points $x^{(k)}$ on ∂S_*) for

$$0 \leq r \leq 0.97, \quad 0 \leq \theta < 2\pi,$$

together with those of the components of $\mathscr{D}$, are shown in Fig. 3.1.

The reason for the restriction $r \leq 0.97$ is the increasing influence of the singularities of $D(x,y)$ and $P(x,y)$ for $x \in S^+$ very close to $y \in \partial S$. This drawback can be mitigated by increasing the floating-point accuracy in the vicinity of ∂S, but it is never completely eliminated.

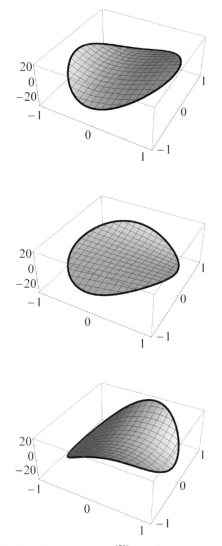

Fig. 3.1 Components of $u^{(51)}$ and $\mathscr{D}$ (heavy lines)

Figure 3.2 contains the graphs of the components of the error $u^{(51)} - u$; the approximation is between 4 and 6 digits of accuracy away from ∂S. These graphs indicate that the accuracy deteriorates significantly close to the boundary, but improves considerably in the interior.

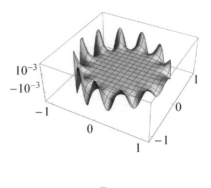

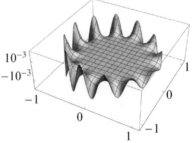

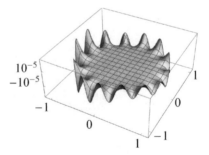

Fig. 3.2 Components of the error $u^{(51)} - u$

3.2.2 Computational Procedure

The step-by-step outline of the computational procedure used for solving the interior Dirichlet problem starts with the Dirichlet boundary data function $\mathscr{D}$, whose components (in polar coordinates) are graphed in Fig. 3.3.

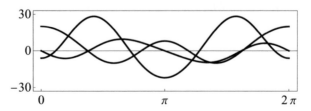

Fig. 3.3 Components of $\mathscr{D}$ in polar coordinates

$\mathscr{D}$ is digitized on ∂S with a sufficient number of points to guarantee that the accuracy of all quadrature calculations significantly exceeds the expected magnitude of the error produced by our approximation method. This allows the method error to be isolated from any quadrature errors. For this example, 100 discrete points on ∂S prove adequate to reach our objective. Next, the functions $\varphi^{(i)}$ are constructed as described in (3.13) and (3.14), and are digitized as well. An orthonormalization method is then selected—in our case, MGS—to generate the digitized $\omega^{(i)}$ as described earlier.

The graphs of typical three-component vector functions $\varphi^{(i)}$ and $\omega^{(i)}$ are displayed in Figs. 3.4 and 3.5, respectively. The increasing number of oscillations in the latter eventually produces computational difficulties in the evaluation of the inner products occurring in the construction of additional orthonormal vector functions.

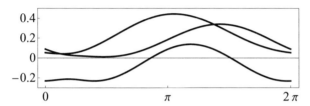

Fig. 3.4 Typical vector functions $\varphi^{(i)}$ in polar coordinates

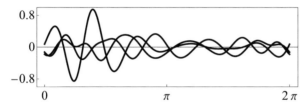

Fig. 3.5 Typical vector functions $\omega^{(i)}$ in polar coordinates

3.2.3 Comparison of CGS, MGS, and HR

The collection of all the inner products $\langle \omega^{(i)}, \omega^{(j)} \rangle$ should produce the identity matrix. The inaccuracy in the orthonormalization process can be seen from the density plot of the difference between the actual $(\langle \omega^{(i)}, \omega^{(j)} \rangle)$ matrix and the identity matrix. This plot, constructed with $\{\omega^{(i)}\}_{i=1}^{99}$ for CGS (see Fig. 3.6), indicates that the method breaks down after about 50 vector functions have been orthonormalized. Here, white means 0 and black means 1 (complete loss of orthogonality).

In correct orthonormalization, the 'angle' between $\omega^{(i)}$ and $\omega^{(j)}$, $i \neq j$, should be $\pi/2$, but its size diminishes when the process deteriorates. The graph of the absolute minimum 'angle' between $\omega^{(i)}$ and $\omega^{(j)}$, $i \neq j$, in the $n \times n$ sub-matrices for $1 \leq n \leq 99$ with CGS is shown in Fig. 3.7. The process breaks down suddenly and severely once inaccuracy takes over.

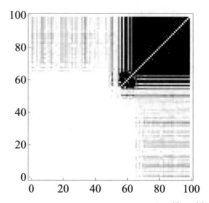

Fig. 3.6 Density plot of the difference between the matrix $(\langle \omega^{(i)}, \omega^{(j)} \rangle)$ and the identity matrix, for $\{\omega^{(i)}\}_{i=1}^{99}$ with CGS

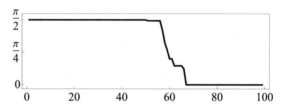

Fig. 3.7 Absolute minimum 'angle' between $\omega^{(i)}$ and $\omega^{(j)}$, $i \neq j$, for $1 \leq n \leq 99$ with CGS

Figure 3.8 illustrates the density plot of the difference between the inner product matrix and the identity matrix for the subset $\{\omega^{(i)}\}_{i=1}^{99}$ with MGS. The orthogonalization is consistently good, with only a slight sign of breaking down.

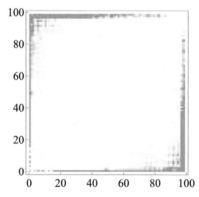

Fig. 3.8 Density plot of the difference between the inner product matrix and the identity matrix, for $\{\omega^{(i)}, i = 1, \ldots, 99\}$ with MGS

The graph of the minimum 'angle' between $\omega^{(i)}$ and $\omega^{(j)}$, $i \neq j$, in the $n \times n$ sub-matrices for $1 \leq n \leq 99$ with MGS is shown in Fig. 3.9. The first signs of deterioration start at about $n = 96$.

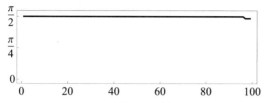

Fig. 3.9 Minimum 'angle' between $\omega^{(i)}$ and $\omega^{(j)}$, $i \neq j$, for $1 \leq n \leq 99$ with MGS

Ill-conditioning increases steadily in the construction of the orthogonal decomposition as computation progresses. Figure 3.10 displays the graph of the relationship in MGS between the condition number and the size of the $n \times n$ matrix R for $n = 1, \ldots, 99$. This graph suggests that the start of the deterioration in the computation may be due to ill-conditioning rather than to the orthogonalization method.

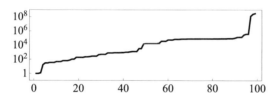

Fig. 3.10 Relationship in MGS between the condition number and the size of the $n \times n$ matrix R for $n = 1, \ldots, 99$

3.13 Remark. The orthonormal vector functions $e^{(k)}$ (in polar coordinates) chosen for HR, which ensure periodicity on $0 \leq \theta < 2\pi$, are

$$\left(\frac{1}{\sqrt{2\pi}}, 0, 0\right)^{\mathsf{T}}, \quad \left(0, \frac{1}{\sqrt{2\pi}}, 0\right)^{\mathsf{T}}, \quad \left(0, 0, \frac{1}{\sqrt{2\pi}}\right)^{\mathsf{T}},$$

$$\left(\frac{\sin\theta}{\sqrt{\pi}}, 0, 0\right)^{\mathsf{T}}, \quad \left(0, \frac{\sin\theta}{\sqrt{\pi}}, 0\right)^{\mathsf{T}}, \quad \left(0, 0, \frac{\sin\theta}{\sqrt{\pi}}\right)^{\mathsf{T}},$$

$$\left(\frac{\cos\theta}{\sqrt{\pi}}, 0, 0\right)^{\mathsf{T}}, \quad \left(0, \frac{\cos\theta}{\sqrt{\pi}}, 0\right)^{\mathsf{T}}, \quad \left(0, 0, \frac{\cos\theta}{\sqrt{\pi}}\right)^{\mathsf{T}},$$

$$\left(\frac{\sin(2\theta)}{\sqrt{\pi}}, 0, 0\right)^{\mathsf{T}}, \quad \left(0, \frac{\sin(2\theta)}{\sqrt{\pi}}, 0\right)^{\mathsf{T}}, \quad \left(0, 0, \frac{\sin(2\theta)}{\sqrt{\pi}}\right)^{\mathsf{T}},$$

$$\left(\frac{\cos(2\theta)}{\sqrt{\pi}}, 0, 0\right)^{\mathsf{T}}, \quad \left(0, \frac{\cos(2\theta)}{\sqrt{\pi}}, 0\right)^{\mathsf{T}}, \quad \left(0, 0, \frac{\cos(2\theta)}{\sqrt{\pi}}\right)^{\mathsf{T}},$$

$$\vdots \qquad\qquad \vdots \qquad\qquad \vdots$$

3.14 Remark. The approximation $u^{(51)}$ computed from $\psi^{(51)}$ by means of HR is virtually indistinguishable from that computed with MGS, which is to be expected since both use the same definite integrals and numerical integration procedures.

The density plot of the difference between the inner product matrix and the identity matrix for the subset $\{\omega^{(i)}\}_{i=1}^{99}$ in HR is shown in Fig. 3.11.

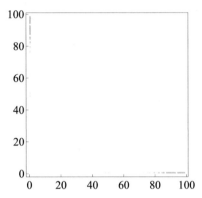

Fig. 3.11 Difference between the inner product matrix and the identity matrix for $i = 1,\ldots,99$ in HR

Finally, Fig. 3.12 shows the graph of the minimum 'angle' between $\omega^{(i)}$ and $\omega^{(j)}$, $i \neq j$, in the $n \times n$ sub-matrices for $1 \leq n \leq 99$ computed with HR. This angle, which should be close to $\pi/2$, shows no significant deterioration.

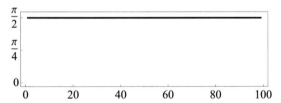

Fig. 3.12 Minimum 'angle' between $\omega^{(i)}$ and $\omega^{(j)}$, $i \neq j$, for $1 \leq n \leq 99$ in HR

3.2.4 Computational Results

For the interior Dirichlet problem, the approximation $\psi^{(n)}$ is computed by means of the system of equations generated by (3.3) and the $x^{(k)}$.

In our example, the approximation $\psi^{(51)}$ of Tu on ∂S was computed using 16 points $x^{(k)}$ on ∂S_* (see Fig. 3.13). As Fig. 3.14 shows, this approximation is quite good when compared with the exact result for Tu derived from the known solution u.

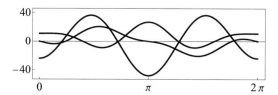

Fig. 3.13 Components of $\psi^{(51)}$ in polar coordinates

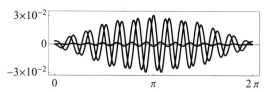

Fig. 3.14 Components of the error $\psi^{(51)} - Tu$ on ∂S in polar coordinates

Equation (3.2) and $\psi^{(51)}$ generate the results for $u^{(51)}$ displayed in Figs. 3.1 and 3.2.

3.2.5 Error Analysis

The approximation $\psi^{(n)}$, $n = 3N + 3$, is computed using N equally spaced points $x^{(k)}$ on ∂S_*. The behavior of the relative error

$$\frac{\|\psi^{(3N+3)} - Tu\|}{\|Tu\|}$$

as a function of N determines the efficiency and accuracy of the computational procedure. This behavior is illustrated by the logarithmic plot in Fig. 3.15.

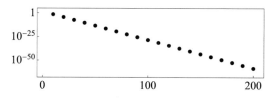

Fig. 3.15 The relative error $\|\psi^{(3N+3)} - Tu\|/\|Tu\|$ for $N = 5, 10, \ldots, 200$

The plot strongly suggests that the relative error improves exponentially as N increases.

Fitting a linear curve to the logarithmic data produces the model

$$1.65658 - 0.300859N.$$

A comparison of the logarithmic data and the linear model is shown in Fig. 3.16. The two are seen to be in good agreement; consequently, the relative error may be modeled by

$$\frac{\|\psi^{(3N+3)} - Tu\|}{\|Tu\|} = 45.3506 \times 10^{-0.300859N}$$

$$= 45.3506 \times 0.500197^{N}.$$

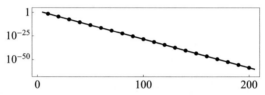

Fig. 3.16 The relative error $\|\psi^{(3N+3)} - Tu\|/\|Tu\|$ and the linear model for $N = 5, 10, \ldots, 200$

The exponentially decreasing error is optimal in the sense that its rate of improvement will eventually exceed that of any fixed-order method. Our model indicates that each additional point $x^{(k)}$ added on ∂S_* will reduce the relative error by about 50%. Clearly, these numbers are specific to our example and should not to be interpreted as universally valid.

3.3 Numerical Example 2

With the same physical plate parameters, auxiliary circle ∂S_*, and points $x^{(k)}$ as in Sect. 3.2, we choose the boundary condition, written in polar coordinates, in (3.1) to be

$$\mathscr{D}(x_p) = \begin{pmatrix} 3 - 4\cos\theta + 4\sin\theta + 4\cos(2\theta) + \cos(4\theta) \\ -1 + 4\cos\theta + 4\sin\theta - 10\sin(2\theta) + \sin(4\theta) \\ 3 - 24\cos\theta + \sin\theta + 2\cos(2\theta) - 2\sin(2\theta) + 10\cos(3\theta) \end{pmatrix}. \quad (3.32)$$

The exact solution u of the problem is not known in this case.

The details mentioned in Remarks 3.9–3.12 apply here as well.

We approximate the solution of (3.1) in this example by means of the MRR method.

3.3.1 Graphical Illustrations

The graphs of the components of $u^{(51)}$ computed from $\psi^{(51)}$ (that is, with 16 points $x^{(k)}$ on ∂S_*) for

$$0 \leq r \leq 0.97, \quad 0 \leq \theta < 2\pi,$$

together with those of the components of $\mathscr{D}$, are shown in Fig. 3.17.

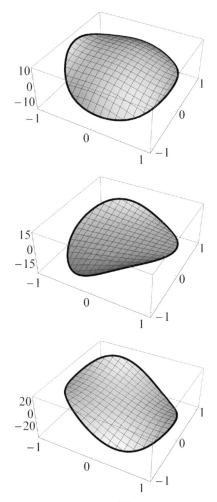

Fig. 3.17 Components of $u^{(51)}$ and $\mathscr{D}$ (heavy lines)

3.3.2 Computational Procedure

The step-by-step outline of the computational procedure used for solving the interior Dirichlet problem starts with the boundary data function $\mathcal{D}$, whose components (in polar coordinates) are graphed in Fig. 3.18.

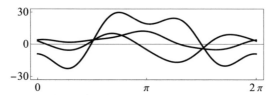

Fig. 3.18 Components of $\mathcal{D}$ in polar coordinates

$\mathcal{D}$ is now digitized on ∂S and the functions $\varphi^{(i)}$ are constructed and digitized as described in Sect. 3.2.2.

3.3.3 Computational Results

For the interior Dirichlet problem, the approximation $\psi^{(n)}$ of Tu is computed by means of the system of equations generated by (3.3) and the $x^{(k)}$. In our example, $\psi^{(51)}$ was calculated with 16 points $x^{(k)}$ on ∂S_* (see Fig. 3.19). This approximation was then combined with the boundary condition (3.32) to generate the graphs in Fig. 3.17, which show good agreement between $u^{(51)}$ and the data function $\mathcal{D}$. This is an indirect validation of our method.

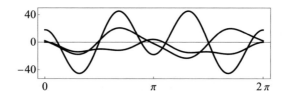

Fig. 3.19 Components of $\psi^{(51)}$ in polar coordinates

3.3.4 Error Analysis

The approximation $\psi^{(n)}$, $n = 3N + 3$, is computed using N equally spaced points $x^{(k)}$ on ∂S_*. The behavior of the relative error

$$\frac{\|\psi^{(3N+3)} - Tu\|}{\|Tu\|}$$

as a function of N determines the efficiency and accuracy of the computational procedure. However, since the exact expression of Tu is not known here, an accurate error analysis cannot be performed.

As a substitute, we propose instead an indirect route to obtaining a measure of partial validation of our calculations. We use the approximation $\psi^{(51)}$ of Tu on ∂S as the boundary condition for an interior Neumann problem and compute the approximate boundary value $(\bar{u}|_{\partial S})^{(51)}$ of the solution to the latter by means of the scheme designed in Chap. 4. If our method is sound, then $(\bar{u}|_{\partial S})^{(51)}$ should be close to $\mathscr{D}$ in (3.32) from which the projection on ∂S of any rigid displacement contained in the solution has been eliminated. In our case, this projection was found to be

$$f = 10f^{(1)} - f^{(2)} + 3f^{(3)}.$$

Figure 3.20 shows the components of $(\bar{u}|_{\partial S})^{(51)} + 10f^{(1)} - f^{(2)} + 3f^{(3)}$.

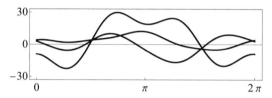

Fig. 3.20 Components of $(\bar{u}|_{\partial S})^{(51)} + 10f^{(1)} - f^{(2)} + 3f^{(3)}$ in polar coordinates

The components of the error $(\bar{u}|_{\partial S})^{(51)} + 10f^{(1)} - f^{(2)} + 3f^{(3)} - \mathscr{D}$ are shown in Fig. 3.21.

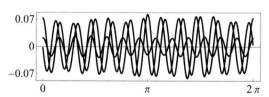

Fig. 3.21 Components of $(\bar{u}|_{\partial S})^{(51)} + 10f^{(1)} - f^{(2)} + 3f^{(3)} - \mathscr{D}$ in polar coordinates

This plot indicates that the total combined error in our verification exercise is consistent with the error expected when we use MRR to compute $(u|_{\partial S})^{(51)}$. Repeating our numerical experiment with 100 and then 200 points $x^{(k)}$ instead of 16 on ∂S_*, we obtained the much smaller relative errors

$$\frac{\|(\bar{u}|_{\partial S})^{(303)} + 10f^{(1)} - f^{(2)} + 3f^{(3)} - \mathscr{D}\|}{\|\mathscr{D}\|} = 2.33551 \times 10^{-27},$$

$$\frac{\|(\bar{u}|_{\partial S})^{(603)} + 10f^{(1)} - f^{(2)} + 3f^{(3)} - \mathscr{D}\|}{\|\mathscr{D}\|} = 1.30003 \times 10^{-55}.$$

This confirms the robust nature of the generalized Fourier series method of approximation presented here.

Chapter 4
Interior Neumann Problem

4.1 Computational Algorithm

The interior Neumann problem consists of the equation and boundary condition

$$Au(x) = 0, \quad x \in S^+,$$
$$Tu(x) = \mathcal{N}(x), \quad x \in \partial S, \tag{4.1}$$

where the vector function $\mathcal{N} \in C^{(0,\alpha)}(\partial S)$, $\alpha \subset (0,1)$, is prescribed.

In this case, the representation formula (1.22) yields the equalities

$$u(x) = \int_{\partial S} D(x,y)\mathcal{N}(y)\,ds(y) - \int_{\partial S} P(x,y)u(y)\,ds(y), \quad x \in S^+, \tag{4.2}$$

$$0 = \int_{\partial S} D(x,y)\mathcal{N}(y)\,ds(y) - \int_{\partial S} P(x,y)u(y)\,ds(y), \quad x \in S^-. \tag{4.3}$$

Let ∂S_* be a simple, closed, C^2–curve surrounding $\bar{S}^+$, let $\{x^{(k)}\}_{k=1}^{\infty}$ be a set of points densely distributed on ∂S_*, and consider the sequence of vector functions on ∂S defined by

$$\mathscr{G} = \{f^{(i)}, \ i = 1,2,3\} \cup \{\varphi^{(jk)}, \ j = 1,2,3, \ k = 1,2,\ldots\},$$

where

$$\varphi^{(jk)}(x) = TD^{(j)}(x,x^{(k)}), \quad j = 1,2,3, \ k = 1,2,\ldots. \tag{4.4}$$

4.1 Remark. The functions $\varphi^{(ij)}$ have an alternative expression, in terms of the rows of P. By (1.10),

$$T(\partial_y)D(y,x) = P^{\mathsf{T}}(x,y),$$

so

$$T(\partial_y)D^{(j)}(y,x) = (P^{\mathsf{T}}(x,y))^{(j)} = (P_{(j)})^{\mathsf{T}}(x,y),$$

from which

$$\varphi^{(jk)}(x) = TD^{(j)}(x,x^{(k)}) = (P_{(j)})^{\mathsf{T}}(x^{(k)},x). \tag{4.5}$$

© Springer Nature Switzerland AG 2020

C. Constanda, D. Doty, *The Generalized Fourier Series Method*, Developments in Mathematics 65, https://doi.org/10.1007/978-3-030-55849-9_4

4.2 Theorem. *The set $\mathscr{G}$ is linearly independent on ∂S and complete in $L^2(\partial S)$.*

Proof. Assuming the opposite, suppose that there are a positive integer N and real numbers c_i and c_{jk}, $i, j = 1, 2, 3, k = 1, 2, \ldots, N$, not all zero, such that

$$\sum_{i=1}^{3} c_i f^{(i)}(x) + \sum_{j=1}^{3}\sum_{k=1}^{N} c_{jk} \varphi^{(jk)}(x) = 0, \quad x \in \partial S. \tag{4.6}$$

We set

$$g(x) = \sum_{j=1}^{3}\sum_{k=1}^{N} c_{jk} D^{(j)}(x, x^{(k)}), \quad x \in S^+, \tag{4.7}$$

and see that, by (4.4), (4.6), and Theorem 1.9,

$$Ag = 0 \quad \text{in } S^+,$$

$$Tg = -\sum_{i=1}^{3} c_i f^{(i)} \quad \text{on } \partial S;$$

that is, g is a regular solution of the Neumann problem in S^+. Therefore, by (1.23),

$$\langle f^{(j)}, Tg \rangle = -\sum_{i=1}^{3} c_i \langle f^{(j)}, f^{(i)} \rangle = 0, \quad j = 1, 2, 3.$$

Since the $f^{(i)}$ are linearly independent, the 3×3 matrix $(\langle f^{(j)}, f^{(i)} \rangle)$ is nonsingular, so $c_i = 0$, $i = 1, 2, 3$. This means that

$$Tg = 0 \quad \text{on } \partial S.$$

Consequently, g is a regular solution of the homogeneous Neumann problem in S^+; hence, by Theorem 1.29(ii),

$$g = \sum_{i=1}^{3} \beta_i f^{(i)} \quad \text{in } \bar{S}^+$$

for some constants β_i, $i = 1, 2, 3$. Then $g - \sum_{i=1}^{3} \beta_i f^{(i)}$ is the (unique) regular solution of the Dirichlet problem in S^+, which, by (4.7), implies that

$$\bar{g}(x) = \sum_{j=1}^{3}\sum_{k=1}^{N} c_{jk} D^{(j)}(x, x^{(k)}) - \sum_{i=1}^{3} \beta_i f^{(i)}(x) = 0, \quad x \in \bar{S}^+.$$

By analyticity,

$$\bar{g} = 0 \quad \text{in } \bar{S}^+_*,$$

and the linear independence of the set $\mathscr{G}$ on ∂S is established by means of the argument used in the proof of Theorem 3.3.

Suppose now that a function $q \in L^2(\partial S)$ satisfies

$$\langle q, f^{(i)} \rangle = \int_{\partial S} (f^{(i)})^{\mathsf{T}} q \, ds = 0, \quad i = 1, 2, 3, \tag{4.8}$$

$$\langle q, \varphi^{(jk)} \rangle = \int_{\partial S} (\varphi^{(jk)})^{\mathsf{T}} q \, ds = 0, \quad j = 1, 2, 3, \; k = 1, 2, \ldots. \tag{4.9}$$

In view of (4.5), Eq. (4.9) is the same as

$$\int_{\partial S} P(x^{(k)}, x) q(x) \, ds = 0, \quad k = 1, 2, \ldots. \tag{4.10}$$

By Theorem 4.18(i), p. 99 in [15], the double-layer potential with L^2-density

$$(Wq)(x) = \int_{\partial S} P(x, y) q(y) \, ds(y), \quad x \in S^+ \cup S^-,$$

is continuous on ∂S_*, so, since the points $x^{(k)}$, $k = 1, 2, \ldots$, are densely distributed on ∂S_*, from (4.10) we deduce that

$$Wq = 0 \quad \text{on } \partial S_*.$$

Theorem 4.18(ii),(iii), p. 99 in [15] now yields

$$A(Wq) = 0 \quad \text{in } S_*^-,$$
$$Wq = 0 \quad \text{on } \partial S_*,$$
$$Wq \in \mathscr{A},$$

which shows that Wq is the (unique) solution of the homogeneous Dirichlet problem in $\bar{S}_*^-$; hence,

$$Wq = 0 \quad \text{in } \bar{S}_*^-.$$

Since the double-layer potential Wq is analytic in $S^+ \cup S^-$, it follows that

$$Wq = 0 \quad \text{in } S^-.$$

Consequently, by Theorem 4.20, p. 99 in [15],

$$\tfrac{1}{2} q(x) + \int_{\partial S} P(x, y) q(y) \, ds(y) = 0 \quad \text{for a.a. } x \in \partial S,$$

with the integral above understood as principal value. According to Theorem 6.12, p. 145 in [15], $q \in C^{0,\alpha}(\partial S)$ for any $\alpha \in (0, 1)$; therefore, by Theorem 3.22(ii), p. 140 in [14],

$$q = \sum_{i=1}^{3} \gamma_i f^{(i)} \quad \text{on } \partial S$$

for some numbers γ_i. Then, by (4.8),

$$\sum_{i=1}^{3} \gamma_i \langle f^{(i)}, f^{(j)} \rangle = 0, \quad j = 1, 2, 3,$$

and, as earlier in the proof, we conclude that $\gamma_i = 0$, $i = 1, 2, 3$; that is, $q = 0$. The desired result now follows from Theorem 3.1. $\qquad \square$

We re-index the set $\mathscr{G}$ as

$$\{\varphi^{(1)}, \varphi^{(2)}, \varphi^{(3)}, \varphi^{(4)}, \varphi^{(5)}, \varphi^{(5)}, \varphi^{(7)}, \varphi^{(8)}, \varphi^{(9)}, \ldots\},$$

where

$$\varphi^{(i)} = f^{(i)}, \quad i = 1, 2, 3, \tag{4.11}$$

and, for $i = 4, 5, \ldots$,

$$\varphi^{(i)} = \varphi^{(jk)}, \quad i = j + 3k, \ j = 1, 2, 3, \ k = 1, 2, \ldots. \tag{4.12}$$

4.1.1 Row Reduction Method

Our intention is to construct a nonsingular linear system of finitely many algebraic equations to approximate the solution u of (4.1).

Rewriting the representation formulas (4.2) and (4.3) as

$$u(x) = \langle D(x, \cdot), \mathscr{N} \rangle - \langle P(x, \cdot), u \rangle, \quad x \in S^+, \tag{4.13}$$

$$\langle P(x, \cdot), u \rangle = \langle D(x, \cdot), \mathscr{N} \rangle, \quad x \in S^-, \tag{4.14}$$

we aim to use (4.14) to approximate $u|_{\partial S}$, and then to approximate u in S^+ by means of (4.13). Since, according to Theorem 1.34 and Remark 1.35, the null space of the operators defined by the two sides in (4.14) is $\mathscr{F}|_{\partial S}$, before we start the procedure, we need to replace these operators by another pair with a zero null space; in other words, we need to eliminate $f^{(i)}|_{\partial S}$ from the formulation of that equality. The obvious way to do this is to seek the solution u of (4.1) in the form

$$u = u^{(c)} + Fa \ \text{in} \ S^+, \tag{4.15}$$

where

$$Fa = \sum_{i=1}^{3} a_i f^{(i)}, \quad a_i = \text{const}, \tag{4.16}$$

and the 'clean' component $u^{(c)}$ is such that $u^{(c)}|_{\partial S}$ is orthogonal to the space of rigid displacements on ∂S; that is,

$$\langle u^{(c)}, f^{(i)} \rangle = 0, \quad i = 1, 2, 3. \tag{4.17}$$

Since $AFc = 0$ and $TFc = 0$, the vector function $u^{(c)}$ is obviously a solution of problem (4.1).

4.3 Remark. Conditions (4.17) do not necessarily imply that $u^{(c)}$ is free of rigid displacements as a function in S^+. They only guarantee that $u^{(c)}$ is the (unique) solution of problem (4.1) whose boundary trace has no component in $\mathscr{F}|_{\partial S}$.

In view of (1.26), we now have

$$\langle P(x,\cdot), u \rangle = \langle P(x,\cdot), u^{(c)} \rangle + \langle P(x,\cdot), Fc \rangle = \langle P(x,\cdot), u^{(c)} \rangle, \quad x \in S^-. \tag{4.18}$$

We set

$$\psi = u^{(c)}|_{\partial S}.$$

Since the $x^{(k)}$ are points in S^-, from (4.18) and (4.14) it follows that

$$\langle P(x^{(k)},\cdot), \psi \rangle = \langle D(x^{(k)},\cdot), \mathscr{N} \rangle,$$

which, by (4.12) and (4.5), leads to

$$\langle \varphi^{(i)}, \psi \rangle = \langle \varphi^{(jk)}, \psi \rangle = \langle (P_{(j)})^{\mathsf{T}}(x^{(k)},\cdot), \psi \rangle$$
$$= \langle (D_{(j)})^{\mathsf{T}}(x^{(k)},\cdot), \mathscr{N} \rangle, \quad j = 1,2,3, \ k = 1,2,\ldots, \ i = 4,5,\ldots. \tag{4.19}$$

In view of Theorem 4.2, there is a unique expansion

$$\psi = \sum_{h=1}^{\infty} c_h \varphi^{(h)}, \tag{4.20}$$

so we can seek an approximation of ψ of the form

$$\psi^{(n)} = \left(u^{(c)}|_{\partial S} \right)^{(n)} = \sum_{h=1}^{n} c_h \varphi^{(h)}.$$

Given the property (4.17), this reduces to

$$\psi^{(n)} = \left(u^{(c)}|_{\partial S} \right)^{(n)} = \sum_{h=4}^{n} c_h \varphi^{(h)}. \tag{4.21}$$

4.4 Remark. For symmetry and ease of computation, in the construction of the approximation $\psi^{(n)}$ we use the subsequence with $n = 3N+3$, where N is the number of points $x^{(k)}$ selected on ∂S_*. This choice feeds into the computation, for each value of $k = 1,2,\ldots,N$, all three rows of P involved in the definition (4.5) of the vector functions $\varphi^{(i)}$, $i = 4,5,\ldots,3N+3$, with the re-indexing (4.12).

Replacing (4.21) in (4.19), we arrive at the approximate equality

$$\sum_{h=4}^{n} c_h \langle \varphi^{(i)}, \varphi^{(h)} \rangle = \langle D_{(j)}(x^{(k)},\cdot), \mathscr{N} \rangle,$$
$$j = 1,2,3, \ k = 1,2,\ldots,N, \ i = 4,5,\ldots,n = 3N+3. \tag{4.22}$$

Equations (4.22) form our system for determining the $3N$ coefficients c_h. In expanded form, this system is

$$
\begin{pmatrix}
\langle \varphi^{(4)}, \varphi^{(4)} \rangle & \langle \varphi^{(4)}, \varphi^{(5)} \rangle & \cdots & \langle \varphi^{(4)}, \varphi^{(n)} \rangle \\
\langle \varphi^{(5)}, \varphi^{(4)} \rangle & \langle \varphi^{(5)}, \varphi^{(5)} \rangle & \cdots & \langle \varphi^{(5)}, \varphi^{(n)} \rangle \\
\vdots & \vdots & & \vdots \\
\langle \varphi^{(n)}, \varphi^{(4)} \rangle & \langle \varphi^{(n)}, \varphi^{(5)} \rangle & \cdots & \langle \varphi^{(n)}, \varphi^{(n)} \rangle
\end{pmatrix}
\begin{pmatrix}
c_4 \\ c_5 \\ \vdots \\ c_n
\end{pmatrix}
$$

$$
= \begin{pmatrix}
\langle D_{(1)}(x^{(1)}, \cdot), \mathcal{N} \rangle \\
\langle D_{(2)}(x^{(1)}, \cdot), \mathcal{N} \rangle \\
\vdots \\
\langle D_{(3)}(x^{(N)}, \cdot), \mathcal{N} \rangle
\end{pmatrix}.
\tag{4.23}
$$

Next, the property $TFc = 0$ implies that

$$
Tu = Tu^{(c)} = \mathcal{N}.
$$

Also, by (1.26),

$$
\langle P(x, \cdot), u \rangle = \langle P(x, \cdot), u^{(c)} \rangle + \langle P(x, \cdot), Fc \rangle = \langle P(x, \cdot), u^{(c)} \rangle - Fc, \quad x \in S^+.
$$

This equality and the representation formula (4.13) now yield

$$
\begin{aligned}
u(x) &= \langle D(x, \cdot), \mathcal{N} \rangle - \langle P(x, \cdot), u^{(c)} \rangle - \langle P(x, \cdot), Fc \rangle \\
&= \langle D(x, \cdot), \mathcal{N} \rangle - \langle P(x, \cdot), \psi \rangle + (Fc)(x), \quad x \in S^+,
\end{aligned}
$$

which suggests that, suppressing the arbitrary rigid displacement Fc, we should construct the function

$$
(u^{(c)})^{(n)}(x) = \langle D(x, \cdot), \mathcal{N} \rangle - \langle P(x, \cdot), \psi^{(n)} \rangle, \quad x \in S^+.
\tag{4.24}
$$

4.5 Theorem. *The vector function* $(u^{(c)})^{(n)}$ *defined by (4.24) is an approximation of the solution* $u^{(c)}$ *of problem (4.1) in the sense that* $(u^{(c)})^{(n)} \to u^{(c)}$ *uniformly on any closed subdomain* S' *of* S^+.

The proof is similar to that of Theorem 3.6, with D replaced by P.

4.6 Remark. The coefficients a_i, $i = 1, 2, 3$, in (4.16) remain undetermined, which is to be expected since, according to Theorem 1.29, the solution of the interior Neumann problem is unique up to a rigid displacement. Our numerical computation procedure always generates the (unique) solution $u^{(c)}$, which may contain its own, specific, rigid displacement in S^+.

4.1.2 Orthonormalization Method

Since we intend to apply this technique to the trace $u^{(c)}|_{\partial S}$ of the solution $u^{(c)}$ of (4.1), which, according to (4.17), has no rigid displacement component (on ∂S), we operate only with the set $\{\varphi^{(i)}\}_{i=4}^{n}$ and orthonormalize it as

$$\{\omega^{(i)}\}_{i=4}^{n}, \quad n = 3N + 3.$$

This yields the QR factorization

$$\Phi = \Omega R, \tag{4.25}$$

where the matrices Φ, Ω, and the nonsingular, upper-triangular $3N \times 3N$ matrix R, described in Sect. 2.2, are constructed with the corresponding vector functions labelled from 4 to $n = 3N + 3$.

Using the notation

$$M = (M_{ij}) = (\langle \varphi^{(i)}, \varphi^{(j)} \rangle), \quad 4 \le i, j \le n = 3N + 3,$$

$c = (c_4, c_5, \ldots, c_n)^{\mathsf{T}}$, and β as the column vector of the numbers on the right-hand side of (4.23), we rewrite that system in the form

$$Mc = \beta.$$

Then

$$c = M^{-1}\beta,$$

and the approximation $\psi^{(n)}$ of $u^{(c)}|_{\partial S}$ given by (4.21) is

$$\psi^{(n)} = \Phi c = \Phi M^{-1}\beta = \Omega R M^{-1}\beta.$$

Since the direct computation of M^{-1} is very inefficient, we prefer to avoid it by using the orthonormality of the set Ω. According to our notation,

$$M = (\langle \varphi^{(i)}, \varphi^{(j)} \rangle) = \langle \Phi^{\mathsf{T}}, \Phi \rangle,$$

so, by (4.25),

$$M = \langle R^{\mathsf{T}}\Omega^{\mathsf{T}}, \Omega R \rangle = R^{\mathsf{T}}\langle \Omega^{\mathsf{T}}, \Omega \rangle R = R^{\mathsf{T}} I R = R^{\mathsf{T}} R.$$

Hence,

$$RM^{-1} = R(R^{\mathsf{T}}R)^{-1} = RR^{-1}(R^{\mathsf{T}})^{-1} = (R^{\mathsf{T}})^{-1},$$

which leads to

$$\psi^{(n)} = \Omega (R^{\mathsf{T}})^{-1}\beta. \tag{4.26}$$

R is upper triangular, so $(R^{\mathsf{T}})^{-1} = (\bar{r}_{ih})$ is lower triangular and can be computed by forward substitution. This reduces (4.26) to

$$\psi^{(n)} = \sum_{i=4}^{n} \sum_{h=4}^{i} \bar{r}_{ih} \beta_h \omega^{(i)} = \sum_{i=4}^{n} \sum_{h=4}^{i} \bar{r}_{ih} \langle (D_{(j)})^{\mathsf{T}}(x^{(k)}, \cdot), \mathcal{N} \rangle \omega^{(i)},$$

$$j = 1, 2, 3, \quad k = 1, 2, \dots, N, \quad i = j + 3k = 4, 5, \dots, n = 3N + 3.$$

Once $\psi^{(n)}$ has been computed, we can use (4.24) to construct the approximation $(u^{(c)})^{(n)}$ of the solution $u^{(c)}$ of problem (4.1).

Remark 3.7 applies to this calculation as well.

4.2 Numerical Example 1

Let S^+ be the disk of radius 1 centered at the origin, and suppose that, after rescaling and non-dimensionalization, the physical and geometric parameters of the plate are

$$h = 0.5, \quad \lambda = \mu = 1.$$

We choose the boundary condition, written in polar coordinates on ∂S, in (4.1) to be

$$\mathcal{N}(x_p) = \begin{pmatrix} 15\cos\theta + 4\sin(2\theta) + 9\cos(3\theta) - 6\sin(3\theta) \\ 6 - 15\sin\theta - 4\cos(2\theta) - 6\cos(3\theta) + 9\sin(3\theta) \\ 12\sin\theta + 54\cos(2\theta) \end{pmatrix}. \tag{4.27}$$

It is easily checked that the function in (4.27) satisfies the solvability condition (1.23), and that the exact (unique) solution of problem (4.1) satisfying conditions (4.17) is

$$u^{(c)}(x) = \begin{pmatrix} -6x_1 + 4x_1x_2 + 12x_1^3 - 12x_1^2x_2 + 4x_2^3 \\ \frac{2}{3} + 6x_2 + 2x_1^2 + 6x_2^2 - 4x_1^3 + 12x_1x_2^2 - 12x_2^3 \\ \frac{34}{3}x2 + 30x_1^2 - 30x_2^2 - 2x_1^2x_2 - 2x_2^3 - 3x_1^4 + 4x_1^3x_2 - 4x_1x_2^3 + 3x_2^4 \end{pmatrix}.$$

Applying the procedure set out in Remark 1.35, we find that this solution contains the rigid displacement (see Remark 4.3)

$$Fa = -\frac{34}{3} f^{(2)} \quad \text{in } S^+.$$

We select the auxiliary curve ∂S_* to be the circle of radius 2 centered at the origin.

The details mentioned in Remarks 3.9–3.12 apply here as well.

We approximate $u^{(c)}$ in this example by means of the MGS method.

4.2.1 Graphical Illustrations

The graphs of the components of $(u^{(c)})^{(51)}$ computed from $\psi^{(51)}$ (that is, with 16 points $x^{(k)}$ on ∂S_*) for

$$0 \le r \le 0.97, \quad 0 \le \theta < 2\pi,$$

and those of the components of $\psi^{(51)} = (u^{(c)}|_{\partial S})^{(51)}$, are shown in Fig. 4.1.

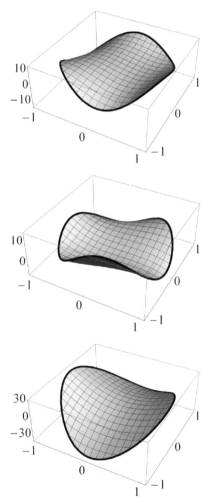

Fig. 4.1 Components of $(u^{(c)})^{(51)}$ and $\psi^{(51)} = (u^{(c)}|_{\partial S})^{(51)}$ (heavy lines)

The reason for the restriction $r \le 0.97$ is the increasing influence of the singularities of $D(x,y)$ and $P(x,y)$ for $x \in S^+$ very close to $y \in \partial S$. This drawback can be mitigated by increasing the floating-point accuracy in the vicinity of ∂S, but it is never completely eliminated.

Figure 4.2 contains the graphs of the components of the error $(u^{(c)})^{(51)} - u^{(c)}$; the approximation is between 4 and 6 digits of accuracy away from ∂S. These graphs indicate that the accuracy deteriorates significantly close to the boundary, but improves considerably in the interior.

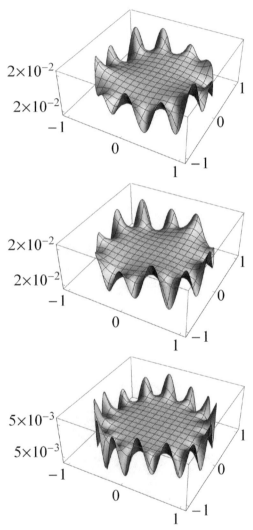

Fig. 4.2 Components of the error $(u^{(c)})^{(51)} - u^{(c)}$

4.2.2 Computational Procedure

The step-by-step outline of the computational procedure used for solving the interior Neumann problem starts with the boundary data function $\mathcal{N}$, whose components (in polar coordinates) are graphed in Fig. 4.3.

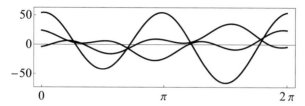

Fig. 4.3 Components of $\mathcal{N}$ in polar coordinates

$\mathcal{N}$ is digitized on ∂S with a sufficient number of points to guarantee that the accuracy of all quadrature calculations significantly exceeds the expected magnitude of the error produced by our approximation method. This allows the method error to be isolated from any quadrature errors. For this example, 100 discrete points on ∂S prove adequate to reach our objective. Next, the functions $\varphi^{(i)}$ are constructed as described in (4.4) and (4.12), and are digitized as well. An orthonormalization method is then selected—in our case, MGS   to generate the digitized $\omega^{(i)}$ as described earlier.

The graphs of typical three-component vector functions $\varphi^{(i)}$ and $\omega^{(i)}$ are displayed in Figs. 4.4 and 4.5, respectively. The increasing number of oscillations in the latter eventually produces computational difficulties in the evaluation of the inner products occurring in the construction of additional orthonormal vector functions.

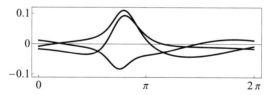

Fig. 4.4 Typical vector functions $\varphi^{(i)}$ in polar coordinates

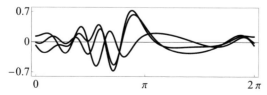

Fig. 4.5 Typical vector functions $\omega^{(i)}$ in polar coordinates

4.2.3 Computational Results

For the interior Neumann problem, the approximation $\psi^{(n)}$ of $u^{(c)}|_{\partial S}$ is computed by means of the system of equations generated by (4.19) and the $x^{(k)}$.

In our example, $\psi^{(51)}$ was calculated with 16 points $x^{(k)}$ on ∂S_* (see Fig. 4.6). As Fig. 4.7 shows, this approximation is quite good when compared with the exact result for $u^{(c)}|_{\partial S}$ obtained from the known solution $u^{(c)}$.

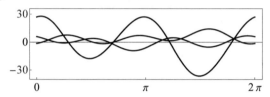

Fig. 4.6 Components of $\psi^{(51)}$ in polar coordinates

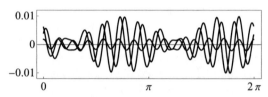

Fig. 4.7 Components of the error $\psi^{(51)} - u^{(c)}|_{\partial S}$ in polar coordinates

Equation (4.13) and $\psi^{(51)}$ generate the results displayed in Figs. 4.1 and 4.2.

4.2.4 Error Analysis

The approximation $\psi^{(n)}$, $n = 3N + 3$, is computed using N equally spaced points $x^{(k)}$ on ∂S_*. The behavior of the relative error

$$\frac{\|\psi^{(3N+3)} - u^{(c)}|_{\partial S}\|}{\|u^{(c)}|_{\partial S}\|}$$

as a function of N, illustrated by the logarithmic plot in Fig. 4.8, determines the efficiency and accuracy of the computational procedure.

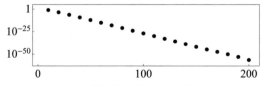

Fig. 4.8 The relative error $\|\psi^{(3N+3)} - u^{(c)}|_{\partial S}\|/\|u^{(c)}|_{\partial S}\|$ for $N = 5, 10, \ldots, 200$

This plot strongly suggests that the relative error improves exponentially as N increases.

Fitting a linear curve to the logarithmic data produces the model

$$2.46579 - 0.2935N.$$

A comparison of the logarithmic data and the linear model is shown in Fig. 4.9. The two are seen to be in good agreement; consequently, the relative error may be modeled by

$$\frac{\|\psi^{(3N+3)} - u^{(c)}|_{\partial S}\|}{\|u^{(c)}|_{\partial S}\|} = 292.276 \times 10^{-0.2935N}$$

$$= 292.276 \times 0.508744^N.$$

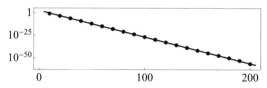

Fig. 4.9 The relative error $\|\psi^{(3N+3)} - u^{(c)}|_{\partial S}\|/\|u^{(c)}|_{\partial S}\|$ together with the linear model for $N = 5, 10, \ldots, 200$

The exponentially decreasing error is optimal in the sense that its rate of improvement will eventually exceed that of any fixed-order method. Our model indicates that each additional point $x^{(k)}$ added on ∂S_* will reduce the relative error by about 50%. Clearly, these numbers are specific to our example and should not to be interpreted as universally valid.

4.3 Numerical Example 2

With the same plate parameters, auxiliary circle ∂S_*, and points $x^{(k)}$ as in Sect. 4.2, we choose the boundary condition, written in polar coordinates, in (4.1) to be

$$\mathcal{N}(x_p) = \begin{pmatrix} -6 + 18\cos\theta - 4\cos(2\theta) + 4\sin(2\theta) + 33\cos(3\theta) \\ -3 - 18\sin\theta + 8\cos(2\theta) - 4\sin(2\theta) - 15\sin(3\theta) \\ -12\cos\theta - 6\sin\theta + 54\cos(2\theta) \end{pmatrix}. \quad (4.28)$$

It is easily checked that the function in (4.28) satisfies the solvability condition (1.23). However, the exact solution $u^{(c)}$ of problem (4.1) is not known in this case.

The details mentioned in Remarks 3.9–3.12 apply here as well.

The computational method of choice in this example is MRR.

4.3.1 Graphical Illustrations

The graphs of the components of $(u^{(c)})^{(51)}$ computed from $\psi^{(51)}$ (that is, with 16 points $x^{(k)}$ on ∂S_*) for

$$0 \le r \le 0.97, \quad 0 \le \theta < 2\pi,$$

together with those of the components of $(u^{(c)})^{(51)}|_{\partial S} = \psi^{(51)}$, are shown in Fig. 4.10. We mention that $\psi^{(51)}$ contains no rigid displacement.

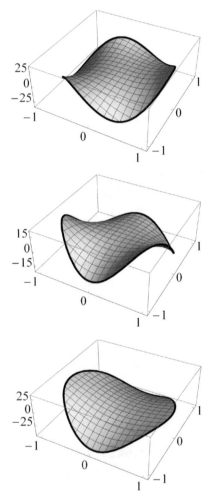

Fig. 4.10 Components of $(u^{(c)})^{(51)}$ and $(u^{(c)}|_{\partial S})^{(51)} = \psi^{(51)}$ (heavy lines)

4.3.2 Computational Procedure

The step-by-step outline of the computational procedure used for solving the interior Neumann problem starts with the boundary data function $\mathcal{N}$, whose components (in polar coordinates) are graphed in Fig. 4.11.

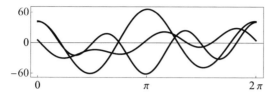

Fig. 4.11 Components of $\mathcal{N}$ in polar coordinates

$\mathcal{N}$ is now digitized on ∂S and the functions $\varphi^{(i)}$ are constructed and digitized as described in Sect. 4.2.2.

4.3.3 Computational Results

For the interior Neumann problem, the approximation $\psi^{(n)}$ of $u^{(c)}|_{\partial S}$ is computed by means of the system of equations generated by (4.19) and the $x^{(k)}$. In our example, $\psi^{(51)}$ was calculated with 16 points $x^{(k)}$ on ∂S_* (see Fig. 4.12). This approximation was then combined with $(u^{(c)})^{(51)}$ to generate the graphs in Fig. 4.10, which show good agreement between $(u^{(c)})^{(51)}$ and $\psi^{(51)}$. This is an indirect validation of our method.

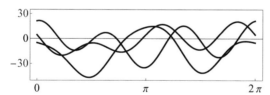

Fig. 4.12 Components of $\psi^{(51)}$ in polar coordinates

4.3.4 Error Analysis

The approximation $\psi^{(n)}$, $n = 3N + 3$, is computed using N equally spaced points $x^{(k)}$ on ∂S_*. The behavior of the relative error

$$\frac{\|\psi^{(3N+3)} - u^{(c)}|_{\partial S}\|}{\|u^{(c)}|_{\partial S}\|}$$

as a function of N determines the efficiency and accuracy of the computational procedure. However, since the exact expression of $u^{(c)}|_{\partial S}$ is not known in this example, an accurate error analysis cannot be performed.

As a substitute, we propose instead an indirect route to obtaining a measure of partial validation of our calculations. We use the approximation $\psi^{(51)}$ of $u^{(c)}|_{\partial S}$ as the boundary condition for an interior Dirichlet problem and compute $T\bar{u}^{(51)}$ for the approximate solution $\bar{u}^{(51)}$ of the latter, by means of the scheme designed in Chap. 3. If our method is sound, then $T\bar{u}^{(51)}$ should be close to $\mathcal{N}$ in (4.28). Figure 4.13 shows the components of $T\bar{u}^{(51)}$.

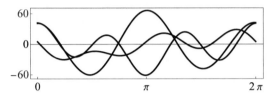

Fig. 4.13 Components of $T\bar{u}^{(51)}$ in polar coordinates

The components of the difference $T\bar{u}^{(51)} - \mathcal{N}$ are shown in Fig. 4.14.

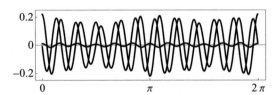

Fig. 4.14 Components of $T\bar{u}^{(51)} - \mathcal{N}$ in polar coordinates

This plot indicates that the total combined error in our verification exercise is consistent with the error expected when we use MRR to compute $(u^{(c)})^{(51)}$. Repeating the numerical experiment with 200 points $x^{(k)}$ instead of 16 on ∂S_*, we obtained the much smaller relative error

$$\frac{\|T\bar{u}^{(603)} - \mathcal{N}\|}{\|\mathcal{N}\|} = 1.76872 \times 10^{-58},$$

which confirms the robust nature of the generalized Fourier series method of approximation presented here.

Chapter 5
Interior Robin Problem

5.1 Computational Algorithm

The interior Robin problem consists in finding u such that

$$Au(x) = 0, \quad x \in S^+,$$
$$(T + \sigma(x))u(x) = \mathscr{R}(x), \quad x \in \partial S, \tag{5.1}$$

where the symmetric, positive definite, continuous 3×3 matrix function σ and the vector function $\mathscr{R} \in C^{(0,\alpha)}(\partial S)$, $\alpha \in (0,1)$, are prescribed.

Since

$$Tu = \mathscr{R} - \sigma u|_{\partial S},$$

the representation formulas (1.12) can be written as

$$u(x) = -\int_{\partial S} D(x,y)(\sigma u)(y)\,ds(y) - \int_{\partial S} P(x,y)u(y)\,ds(y)$$

$$+ \int_{\partial S} D(x,y)\mathscr{R}(y)\,ds(y), \quad x \in S^+, \tag{5.2}$$

$$0 = -\int_{\partial S} D(x,y)(\sigma u)(y)\,ds(y) - \int_{\partial S} P(x,y)u(y)\,ds(y)$$

$$+ \int_{\partial S} D(x,y)\mathscr{R}(y)\,ds(y), \quad x \in S^-. \tag{5.3}$$

Let ∂S_* be a simple, closed, C^2-curve surrounding $\bar{S}^+$, and let $\{x^{(k)}\}_{k=1}^{\infty}$ be a set of points densely distributed on ∂S_*. We consider the collection of vector functions on ∂S defined by

$$\mathscr{G} = \{\sigma f^{(i)}, \ i = 1,2,3\} \cup \{\varphi^{(jk)}, \ j = 1,2,3, \ k = 1,2,\dots\},$$

© Springer Nature Switzerland AG 2020
C. Constanda, D. Doty, *The Generalized Fourier Series Method*, Developments in
Mathematics 65, https://doi.org/10.1007/978-3-030-55849-9_5

where

$$\varphi^{(jk)}(x) = (T+\sigma(x))D^{(j)}(x,x^{(k)}), \quad j=1,2,3, \ k=1,2,\dots.$$

5.1 Remark. There are two useful alternative expressions for $\varphi^{(jk)}$, in terms of the rows of matrix D and those of matrix P.

(i) By (1.9), and indicating in detail the sequence of operations, we have

$$(T+\sigma(x))D^{(j)}(y,x) = (T+\sigma(x))(D(y,x))^{(j)} = (T+\sigma(x))\left((D(x,y))^{\mathsf{T}}\right)^{(j)}$$
$$= (T+\sigma(x))\left((D(x,y))_{(j)}\right)^{\mathsf{T}} = (T+\sigma(x))(D_{(j)})^{\mathsf{T}}(x,y).$$

With x replaced by $x^{(k)}$ and y replaced by x, this yields

$$\varphi^{(jk)}(x) = (T+\sigma(x))D^{(j)}(x,x^{(k)})$$
$$= (T+\sigma(x))(D_{(j)})^{\mathsf{T}}(x^{(k)},x), \quad j=1,2,3, \ k=1,2,\dots. \tag{5.4}$$

(ii) We rewrite (5.4) as

$$\varphi^{(jk)}(x) = T(D_{(j)})^{\mathsf{T}}(x^{(k)},x) + \sigma(x)(D_{(j)})^{\mathsf{T}}(x^{(k)},x). \tag{5.5}$$

By the definition of P,

$$P_{(j)}(x,y) = \left((T(\partial_y)D(y,x))^{\mathsf{T}}\right)_{(j)} = \left((T(\partial_y)D(y,x))^{(j)}\right)^{\mathsf{T}},$$

so

$$(P_{(j)})^{\mathsf{T}}(x,y) = (T(\partial_y)D(y,x))^{(j)} = (T(\partial_y)D^{\mathsf{T}}(x,y))^{(j)}$$
$$= T(\partial_y)(D^{\mathsf{T}})^{(j)}(x,y) = T(\partial_y)(D_{(j)})^{\mathsf{T}}(x,y).$$

With x and y replaced by $x^{(k)}$ and x, respectively, this becomes

$$(P_j)^{\mathsf{T}}(x^{(k)},x) = T(D_{(j)})^{\mathsf{T}}(x^{(k)},x). \tag{5.6}$$

At the same time,

$$\sigma(x)(D_{(j)})^{\mathsf{T}}(x^{(k)},x) = \sigma^{\mathsf{T}}(x)(D_{(j)})^{\mathsf{T}}(x^{(k)},x) = (D_{(j)}(x^{(k)},x)\sigma(x))^{\mathsf{T}}. \tag{5.7}$$

Combining (5.6) and (5.7) with (5.5), we arrive at

$$\varphi^{(ij)}(x) = (P_{(j)}(x^{(k)},x) + D_{(j)}(x^{(k)},x)\sigma(x))^{\mathsf{T}}, \quad x \in \partial S. \tag{5.8}$$

5.2 Remark. The symmetry of σ produces a useful equality when inner product notation is employed. Thus, if $v,w \in L^2(\partial S)$ are 3×1 column vector functions, then, by Definition 1.2(ii),

$$\langle \sigma v, w \rangle = \int_{\partial S} (\sigma v)^\mathsf{T} w \, ds = \int_{\partial S} v^\mathsf{T} \sigma^\mathsf{T} w \, ds = \int_{\partial S} v^\mathsf{T} \sigma w \, ds = \langle v, \sigma w \rangle. \qquad (5.9)$$

5.3 Theorem. *The set $\mathscr{G}$ is linearly independent on ∂S and complete in $L^2(\partial S)$.*

Proof. Assuming the opposite, suppose that there are a positive integer N and real numbers c_i and c_{jk}, $i, j = 1, 2, 3$, $k = 1, 2, \ldots$, not all zero, such that

$$\sum_{i=1}^{3} c_i \sigma(x) f^{(i)}(x) + \sum_{j=1}^{3} \sum_{k=1}^{N} c_{jk} \varphi^{(jk)}(x) = 0, \quad x \in \partial S. \qquad (5.10)$$

Setting

$$g(x) = \sum_{i=1}^{3} c_i f^{(i)}(x) + \sum_{j=1}^{3} \sum_{k=1}^{N} c_{jk} D^{(j)}(x, x^{(k)}), \quad x \in S^+, \qquad (5.11)$$

and using Theorem 1.9, (5.4), (5.11), (5.10), and the fact that $Af^{(i)} = 0$ in S^+ and $Tf^{(i)} = 0$ on ∂S, we see that g satisfies

$$Ag = 0 \quad \text{in } S^+,$$
$$(T + \sigma)g = 0 \quad \text{on } \partial S;$$

in other words, g is the (unique) regular solution of the homogeneous Robin problem in S^+. According to Theorem 1.29(iii),

$$g = 0 \quad \text{in } \bar{S}^+,$$

so, by analyticity,

$$g = 0 \quad \text{in } \bar{S}^+_*.$$

The linear independence argument is now developed as in the proof of Theorem 3.3, based on the definition (5.11) of g.

Suppose now that a function $q \in L^2(\partial S)$ satisfies

$$\langle q, \sigma f^{(i)} \rangle = \int_{\partial S} (\sigma f^{(i)})^\mathsf{T} q \, ds = 0, \quad i = 1, 2, 3, \qquad (5.12)$$

$$\langle q, \varphi^{(jk)} \rangle = \int_{\partial S} (\varphi^{(jk)})^\mathsf{T} q \, ds = 0, \quad j = 1, 2, 3, \ k = 1, 2, \ldots.$$

According to (5.9) with v and w replaced by q and $f^{(i)}$, respectively, and (5.12),

$$\langle \sigma q, f^{(i)} \rangle = \langle q, \sigma f^{(i)} \rangle = 0. \qquad (5.13)$$

Next, by Remark 3.2,

$$TD^{(j)}(x, x^{(k)}) = \left(TD(x, x^{(k)}) \right)^{(j)}$$
$$= \left(P^\mathsf{T}(x^{(k)}, x) \right)^{(j)} = (P_{(j)})^\mathsf{T}(x^{(k)}, x), \qquad (5.14)$$

so

$$\int_{\partial S} \left(TD^{(j)}(x,x^{(k)})\right)^{\mathsf{T}} q(x)\,ds = \int_{\partial S} P_{(j)}(x^{(k)},x)q(x)\,ds.$$

Then the double-layer potential with L^2-density

$$(Wq)(x) = \int_{\partial S} P(x,y)q(y)\,ds(y), \quad x \in S^+ \cup S^-,$$

satisfies

$$(Wq)_j(x^k) = \int_{\partial S} P_{(j)}(x^{(k)},x)q(x)\,ds(x) = \int_{\partial S} \left(TD^{(j)}(x,x^{(k)})\right)^{\mathsf{T}} q(x)\,ds(x). \quad (5.15)$$

Finally, by the symmetry of σ,

$$\begin{aligned}
\left(\sigma(x)D^{(j)}(x,x^{(k)})\right)^{\mathsf{T}} q(x) &= D_{(j)}(x^{(k)},x)\sigma^{\mathsf{T}}(x)q(x) \\
&= D_{(j)}(x^{(k)},x)\sigma(x)q(x),
\end{aligned}$$

so the single-layer potential with L^2-density

$$\left(V(\sigma q)\right)(x) = \int_{\partial S} D(x,y)\sigma(y)q(y)\,ds(y), \quad x \in S^+ \cup S^-,$$

satisfies

$$\begin{aligned}
\left(V(\sigma q)\right)_j(x^{(k)}) &= \int_{\partial S} D_{(j)}(x^{(k)},x)\sigma(x)q(x)\,ds \\
&= \int_{\partial S} \left(\sigma(x)D^{(j)}(x,x^{(k)})\right)^{\mathsf{T}} q(x)\,ds. \quad (5.16)
\end{aligned}$$

From (5.4), (5.15), and (5.16) it now follows that

$$\begin{aligned}
(Wq)_j&(x^{(k)}) + \left(V(\sigma q)\right)_j(x^{(k)}) \\
&= \int_{\partial S} \left(TD^{(j)}(x,x^{(k)})\right)^{\mathsf{T}} q(x)\,ds(x) + \int_{\partial S} \left(\sigma(x)D^{(j)}(x,x^{(k)})\right)^{\mathsf{T}} q(x)\,ds(x) \\
&= \int_{\partial S} \left((T+\sigma(x))D^{(j)}(x,x^{(k)})\right)^{\mathsf{T}} q(x)\,ds(x) \\
&= \int_{\partial S} \left(\varphi^{(jk)}\right)^{\mathsf{T}}(x)q(x)\,ds(x) = 0.
\end{aligned}$$

Since the set $\{x^{(k)}\}_{k=1}^{\infty}$ is dense on ∂S_*, we conclude that

$$(Wq)(x) + (V(\sigma q))(x) = 0, \quad x \in \partial S_*. \tag{5.17}$$

Also, by Theorems 4.18, p. 99 and 4.2, p. 84 in [15], (5.13), and (5.17), the function

$$U(x) = (Wq)(x) + (V(\sigma q))(x), \quad x \in S^+ \cup S^-,$$

satisfies

$$AU = 0 \quad \text{in } S_*^-,$$
$$U = 0 \quad \text{on } \partial S_*,$$
$$U \in \mathscr{A},$$

which, by Theorem 1.29(iv), implies that

$$U = 0 \quad \text{in } \bar{S}_*^-$$

as the (unique) regular solution of the homogeneous Dirichlet problem in S_*^-. By analyticity, we then have

$$U = 0 \quad \text{in } S^-;$$

hence,

$$U^- = 0 \quad \text{on } \partial S, \tag{5.18}$$
$$(TU)^- = 0 \quad \text{on } \partial S. \tag{5.19}$$

From Theorems 4.20, p. 99 and 4.21, p. 100 in [15] and (5.19) it follows that

$$\tfrac{1}{2}q + W_0 q + V_0(\sigma q) = 0 \quad \text{on } \partial S,$$

or, what is the same,

$$\tfrac{1}{2}q(x) + \int\limits_{\partial S} P(x,y)q(y)\,ds(y) + \int\limits_{\partial S} D(x,y)\sigma(y)q(y)\,ds(y) = 0, \quad x \in \partial S.$$

This equation is of the form

$$\tfrac{1}{2}q(x) + \int\limits_{\partial S} k(x,y)q(y)\,ds(y) = 0, \quad x \in \partial S,$$

where, according to the properties of the layer potentials with L^2-densities (see [15], Section 4.2),

$$k(x,y) = P(x,y) + D(x,y)\sigma(y)$$

is a so-called α-regular singular kernel on ∂S. Then, repeating the proof of Theorem 6.12, p. 145 in [15], we conclude that $q \in C^{1,\alpha}(\partial S)$ for any $\alpha \in (0,1)$, so we can use the properties of V and W in Theorem 1.33 to compute

$$\begin{aligned}
(TU)^+ &= \big(T\big(Wq+V(\sigma q)\big)\big)^+ \\
&= (T(Wq))^+ + (TV(\sigma q))^+ \\
&= (T(Wq))^+ + \tfrac{1}{2}\sigma q + W_0^*(\sigma q) \\
&= (T(Wq))^- + \tfrac{1}{2}\sigma q + \tfrac{1}{2}\sigma q + (TV(\sigma q))^- \\
&= (T(Wq))^- + (TV(\sigma q))^- + \sigma q \\
&= \big(T\big(Wq+V(\sigma q)\big)\big)^- + \sigma q = (TU)^- + \sigma q
\end{aligned}$$

and

$$\begin{aligned}
(\sigma U)^+ &= \sigma\big[(Wq)^+ + (V(\sigma q))^+\big] \\
&= \sigma\big[-\tfrac{1}{2}q + W_0 q + (V(\sigma q))^+\big] \\
&= \sigma\big[-\tfrac{1}{2}q + \big(-\tfrac{1}{2}q + (Wq)^-\big) + (V(\sigma q))^-\big] \\
&= \sigma(-q + U^-) = -\sigma q + \sigma U^+.
\end{aligned}$$

Hence, by (5.18) and (5.19),

$$\big((T+\sigma)U\big)^+ = \big((T+\sigma)U\big)^- = 0 \quad \text{on } \partial S,$$

so U is the (unique) regular solution of the homogeneous Robin problem

$$AU = 0 \quad \text{in } S^+,$$
$$\big((T+\sigma)U\big)^+ = 0 \quad \text{on } \partial S;$$

consequently, by Theorem 1.29(iii),

$$U = 0 \quad \text{in } S^+.$$

Then, taking (5.19) into account, we find that

$$(TU)^+ = (TU)^- = 0 \quad \text{on } \partial S.$$

The first equality above translates as

$$(T(Wq))^+ + (TV(\sigma q))^+ = (T(Wq))^- + (TV(\sigma q))^- \quad \text{on } \partial S.$$

Using the fact that $(T(Wq))^+ = (T(Wq))^-$ and the expressions for $(TV)^+$ and $(TV)^-$ in Theorem 1.33(iii), we arrive at

$$\tfrac{1}{2}\sigma q + W_0^*(\sigma q) = -\tfrac{1}{2}\sigma q + W_0^*(\sigma q) \quad \text{on } \partial S,$$

from which $\sigma q = 0$. Since σ is positive definite, this implies that $q = 0$, thus proving that, according to Theorem 3.1, $\mathscr{G}$ is a complete set in $L^2(\partial S)$. $\square$

We order the elements of $\mathscr{G}$ as the sequence

$$\sigma f^{(1)}, \sigma f^{(2)}, \sigma f^{(3)}, \varphi^{(11)}, \varphi^{(21)}, \varphi^{(31)}, \varphi^{(12)}, \varphi^{(22)}, \varphi^{(32)}, \dots$$

and re-index them, noticing that for each $i = 4, 5, \dots$, there is a unique pair $\{j, k\}$, $j = 1, 2, 3$, $k = 1, 2, \dots$, such that $i = j + 3k$. Then

$$\mathscr{G} = \{\varphi^{(1)}, \varphi^{(2)}, \varphi^{(3)}, \varphi^{(4)}, \varphi^{(5)}, \varphi^{(6)}, \varphi^{(7)}, \varphi^{(8)}, \varphi^{(9)}, \dots\}, \tag{5.20}$$

where

$$\varphi^{(i)} = \sigma f^{(i)}, \quad i = 1, 2, 3, \tag{5.21}$$

and for $i = 4, 5, \dots$,

$$\varphi^{(i)} = \varphi^{(jk)}, \quad i = j + 3k, \; j = 1, 2, 3, \; k = 1, 2, \dots. \tag{5.22}$$

5.1.1 Row Reduction Method

Our intention is to construct a nonsingular linear system of finitely many algebraic equations to approximate the solution u of (5.1).

Rewriting the representation formulas (5.2) and (5.3) as

$$u(x) = -\langle D(x, \cdot)\sigma + P(x, \cdot), u\rangle + \langle D(x, \cdot), \mathscr{R}\rangle, \quad x \in S^+, \tag{5.23}$$

$$\langle D(x, \cdot)\sigma + P(x, \cdot), u\rangle = \langle D(x, \cdot), \mathscr{R}\rangle, \quad x \in S^-, \tag{5.24}$$

we aim to use (5.24) to approximate $u|_{\partial S}$, and then to approximate u in S^+ by means of (5.23).

5.4 Remark. In the interior Dirichlet problem, the rigid displacement in the solution is computed separately. In the interior Neumann problem, it remains completely arbitrary and plays no role in the calculation. In both these cases, the linear algebraic system generated by the boundary data has a nonzero null space that needs special attention. Here, by contrast, the rigid displacement cannot be isolated from the full numerical computation process, which produces a unique solution for $u|_{\partial S}$.

We set

$$\psi = u|_{\partial S}.$$

Since the $x^{(k)}$ are points in S^-, from (5.3) it follows that

$$\int_{\partial S} D(x^{(k)}, x)\sigma(x)\psi(x)\, ds(x) + \int_{\partial S} P(x^{(k)}, x)\psi(x)\, ds(x)$$

$$= \int_{\partial S} D(x^{(k)}, x)\mathscr{R}(x)\, ds(x),$$

which, by (5.22) and (5.8), leads to

$$
\langle \varphi^{(i)}, \psi \rangle = \langle \varphi^{(jk)}, \psi \rangle = \int_{\partial S} (\varphi^{(jk)})^{\mathsf{T}}(x)\psi(x)\,ds(x)
$$

$$
= \int_{\partial S} \left(P_{(j)}(x^{(k)},x) + D_{(j)}(x^{(k)},x)\sigma(x) \right) \psi(x)\,ds(x)
$$

$$
= \int_{\partial S} D_{(j)}(x^{(k)},x)\mathscr{R}(x)\,ds(x), \quad j=1,2,3,\ k=1,2,\dots,\ i=4,5,\dots,
$$

or, what is the same,

$$
\langle \varphi^{(i)}, \psi \rangle = \langle (D_{(j)})^{\mathsf{T}}(x^{(k)}, \cdot), \mathscr{R} \rangle. \tag{5.25}
$$

At the same time, since $Tu = \mathscr{R} - \sigma\psi$ is the Neumann boundary data of the solution, we have

$$
\langle Tu, f^{(i)} \rangle = 0, \quad i=1,2,3,
$$

so

$$
\langle \sigma\psi, f^{(i)} \rangle = \langle \mathscr{R}, f^{(i)} \rangle, \quad i=1,2,3,
$$

which, by (5.9), yields

$$
\langle \psi, \sigma f^{(i)} \rangle = \langle \sigma\psi, f^{(i)} \rangle = \langle \mathscr{R}, f^{(i)} \rangle, \quad i=1,2,3. \tag{5.26}
$$

In view of Theorem 5.3, there is a unique expansion

$$
\psi = \sum_{h=1}^{\infty} c_h \varphi^{(h)},
$$

so we can seek an approximation of ψ of the form

$$
\psi^{(n)} = (u|_{\partial S})^{(n)} = \sum_{h=1}^{n} c_h \varphi^{(h)}. \tag{5.27}
$$

5.5 Remark. For symmetry and ease of computation, in the construction of the approximation $\psi^{(n)}$ we use the subsequence with $n = 3N+3$, where N is the number of points $x^{(k)}$ selected on ∂S_*. This choice feeds into the computation, for each $k=1,2,\dots,N$, all three columns/rows of D involved in the definition (5.4) of the vector functions $\varphi^{(i)}$, $i=4,5,\dots,3N+3$, with the re-indexing (5.22).

Since ψ satisfies (5.26), it is reasonable to ask its approximation $\psi^{(n)}$ to do the same; that is,

$$
\langle \psi^{(n)}, \sigma f^{(i)} \rangle = \langle \mathscr{R}, f^{(i)} \rangle, \quad i=1,2,3. \tag{5.28}
$$

Combining (5.27) with (5.28), and then (5.27) with (5.25), we obtain the approximate equalities

$$\sum_{h=1}^{n} c_h \langle \sigma f^{(i)}, \varphi^{(h)} \rangle = \langle f^{(i)}, \mathscr{R} \rangle, \quad i = 1, 2, 3, \tag{5.29}$$

$$\sum_{h=1}^{n} c_h \langle \varphi^{(i)}, \varphi^{(h)} \rangle = \langle (D_{(j)})^{\mathsf{T}} (x^{(k)}, \cdot), \mathscr{R} \rangle,$$

$$j = 1, 2, 3, \; k = 1, 2, \ldots, N, \; i = j + 3k = 4, 5, \ldots, 3N + 3. \tag{5.30}$$

Equations (5.29) and (5.30) form our system for determining the $3N + 3$ coefficients c_h. In expanded form, this system is

$$\begin{pmatrix} \langle \sigma f^{(1)}, \varphi^{(1)} \rangle & \langle \sigma f^{(1)}, \varphi^{(2)} \rangle & \cdots & \langle \sigma f^{(1)}, \varphi^{(n)} \rangle \\ \langle \sigma f^{(2)}, \varphi^{(1)} \rangle & \langle \sigma f^{(2)}, \varphi^{(2)} \rangle & \cdots & \langle \sigma f^{(2)}, \varphi^{(n)} \rangle \\ \langle \sigma f^{(3)}, \varphi^{(1)} \rangle & \langle \sigma f^{(3)}, \varphi^{(2)} \rangle & \cdots & \langle \sigma f^{(3)}, \varphi^{(n)} \rangle \\ \langle \varphi^{(4)}, \varphi^{(1)} \rangle & \langle \varphi^{(4)}, \varphi^{(2)} \rangle & \cdots & \langle \varphi^{(4)}, \varphi^{(n)} \rangle \\ \langle \varphi^{(5)}, \varphi^{(1)} \rangle & \langle \varphi^{(5)}, \varphi^{(2)} \rangle & \cdots & \langle \varphi^{(5)}, \varphi^{(n)} \rangle \\ \vdots & \vdots & & \vdots \\ \langle \varphi^{(n)}, \varphi^{(1)} \rangle & \langle \varphi^{(n)}, \varphi^{(2)} \rangle & \cdots & \langle \varphi^{(n)}, \varphi^{(n)} \rangle \end{pmatrix} \begin{pmatrix} c_1 \\ c_2 \\ c_3 \\ c_4 \\ c_5 \\ \vdots \\ c_n \end{pmatrix}$$

$$= \begin{pmatrix} \langle f^{(1)}, \mathscr{R} \rangle \\ \langle f^{(2)}, \mathscr{R} \rangle \\ \langle f^{(3)}, \mathscr{R} \rangle \\ \langle D_{(1)}^{\mathsf{T}} (x^{(1)}, \cdot), \mathscr{R} \rangle \\ \langle D_{(2)}^{\mathsf{T}} (x^{(1)}, \cdot), \mathscr{R} \rangle \\ \vdots \\ \langle D_{(3)}^{\mathsf{T}} (x^{(N)}, \cdot), \mathscr{R} \rangle \end{pmatrix}. \tag{5.31}$$

5.6 Remark. We have deliberately written $\sigma f^{(i)}$ instead of $\varphi^{(i)}$ in the inner products in the first three rows in (5.31), to emphasize that those equations have a different source than the rest.

The representation formula (5.23) now suggests that we should construct the function

$$u^{(n)}(x) = -\langle D(x, \cdot) \sigma + P(x, \cdot), \psi^{(n)} \rangle + \langle D(x, \cdot), \mathscr{R} \rangle, \quad x \in S^+. \tag{5.32}$$

5.7 Theorem. *The vector function $u^{(n)}$ defined by (5.32) is an approximation of the solution u of problem (5.1) in the sense that $u^{(n)} \to u$ uniformly on any closed subdomain S' of S^+.*

The proof of this statement is similar to that of Theorem 3.6, with D replaced by $D\sigma + P$.

5.1.2 Orthonormalization Method

After the re-indexing (5.21), (5.22), we orthonormalize the set $\{\varphi^{(i)}\}_{i=1}^n$ as

$$\{\omega^{(i)}\}_{i=1}^n, \quad n = 3N + 3.$$

This yields the QR factorization

$$\Phi = \Omega R, \tag{5.33}$$

where the matrices Φ, Ω, and the nonsingular, upper-triangular $n \times n$ matrix R are described in Sect. 2.2.

Using the notation

$$M = (M_{ij}) = (\langle \varphi^{(i)}, \varphi^{(j)} \rangle), \quad 4 \le i, j \le n = 3N + 3,$$

$c = (c_4, c_5, \ldots, c_n)^\mathsf{T}$, and β as the column vector of the numbers on the right-hand side of (5.31), we rewrite that system in the form

$$Mc = \beta.$$

Then

$$c = M^{-1}\beta,$$

and the approximation $\psi^{(n)}$ of $u^{(c)}|_{\partial S}$ given by (5.27) is

$$\psi^{(n)} = \Phi c = \Phi M^{-1}\beta = \Omega R M^{-1}\beta.$$

Since the direct computation of M^{-1} is very inefficient, we prefer to avoid it by using the orthonormality of the set Ω. According to our notation,

$$M = (\langle \varphi^{(i)}, \varphi^{(j)} \rangle) = \langle \Phi^\mathsf{T}, \Phi \rangle,$$

so, by (5.33),

$$M = \langle R^\mathsf{T}\Omega^\mathsf{T}, \Omega R \rangle = R^\mathsf{T}\langle \Omega^\mathsf{T}, \Omega \rangle R = R^\mathsf{T} I R = R^\mathsf{T} R.$$

Hence,

$$RM^{-1} = R(R^\mathsf{T}R)^{-1} = RR^{-1}(R^\mathsf{T})^{-1} = (R^\mathsf{T})^{-1},$$

which leads to

$$\psi^{(n)} = \Omega(R^\mathsf{T})^{-1}\beta, \quad n = 3N + 3. \tag{5.34}$$

R is upper triangular, so $(R^\mathsf{T})^{-1} = (\bar{r}_{ih})$ is lower triangular and can be computed by forward substitution. This reduces (5.34) to

$$\psi^{(n)} = \sum_{i=1}^n \left(\sum_{h=1}^i \bar{r}_{ih}\beta_h \right) \omega^{(i)}$$

$$= \sum_{i=1}^3 \left(\sum_{h=1}^i \bar{r}_{ih}\langle f^{(h)}, \mathscr{R} \rangle \right) \omega^{(i)} + \sum_{i=4}^n \left(\sum_{h=1}^i \bar{r}_{ih}\langle (D_{(j)})^\mathsf{T}(x^{(k)}, \cdot), \mathscr{R} \rangle \right) \omega^{(i)}.$$

Once $\psi^{(n)}$ has been computed, we can use (5.32) to construct the approximation $u^{(n)}$ of the solution u of problem (5.1).

Remark 3.7 applies to this calculation as well.

5.2 Numerical Example 1

Let S^+ be the disk of radius 1 centered at origin, let the physical and geometric parameters (after rescaling and non-dimensionalization) of the plate be

$$h = 0.5, \quad \lambda = \mu = 1,$$

and let

$$\sigma(x) = \begin{pmatrix} 11 & 2 & 1 \\ 2 & 8 & -2 \\ 1 & -2 & 11 \end{pmatrix}.$$

We choose the boundary condition function $\mathcal{R}$ (in polar coordinates on ∂S) in (5.1) to be

$$\mathcal{R}(x_p) = \begin{pmatrix} 15 - 68\cos\theta + 46\sin\theta + 20\cos(2\theta) + 22\sin(2\theta) \\ \quad + 32\cos(3\theta) + 18\sin(3\theta) - \cos(4\theta) \\ 6 - 14\cos\theta + 68\sin\theta + 16\cos(2\theta) - 16\sin(2\theta) \\ \quad - 16\cos(3\theta) - 45\sin(3\theta) + 2\cos(4\theta) \\ -21 + 144\cos\theta - 12\sin\theta - 82\cos(2\theta) + 230\sin(2\theta) \\ \quad + 7\cos(3\theta) + 12\sin(3\theta) - 11\cos(4\theta) \end{pmatrix}.$$

It is easily verified that the exact solution of problem (5.1) in this case is

$$u(x) = \begin{pmatrix} -3 - 4x_1 - 4x_2 + 6x_1^2 + 2x_2^2 + 12x_1^2x_2 - 12x_1x_2^2 + 4x_2^3 \\ -4x_1 + 4x_2 + 4x_1x_2 + 4x_1^3 - 12x_1^2x_2 + 12x_1x_2^2 + 8x_2^3 \\ -2 + 15x_1 - 7x_1^2 + 40x_1x_2 + 7x_2^2 - 2x_1^3 - 2x_1x_2^2 - 4x_1^2x_2 \\ \quad + 6x_1^2x_2^2 - 4x_1x_2^3 - 2x_2^4 \end{pmatrix}.$$

Applying the procedure described in Remark 1.11, we find that this solution contains the rigid displacement

$$f = -15f^{(1)} - 2f^{(3)}.$$

We select the auxiliary curve ∂S_* to be the circle of radius 2 centered at the origin.

The details mentioned in Remarks 3.9–3.12 apply here as well.

We approximate the solution of problem (5.1) in this example by means of the MGS method.

5.2.1 Graphical Illustrations

The graphs of the components of $u^{(51)}$ computed from $\psi^{(51)}$ (that is, with 16 points $x^{(k)}$ on ∂S_*) for

$$0 \le r \le 0.97, \quad 0 \le \theta < 2\pi,$$

together with those of the components of $\psi^{(51)} = (u|_{\partial S})^{(51)}$, are shown in Fig. 5.1.

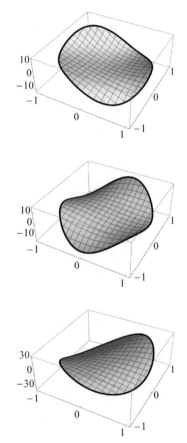

Fig. 5.1 Components of $u^{(51)}$ and $\psi^{(51)} = (u|_{\partial S})^{(51)}$ (heavy lines)

The reason for the restriction $r \le 0.97$ is the increasing influence of the singularities of $D(x, y)$ and $P(x, y)$ for $x \in S^+$ very close to $y \in \partial S$. This drawback can be mitigated by increasing the floating-point accuracy in the vicinity of ∂S, but it is never completely eliminated.

Figure 5.2 contains the graphs of the components of the error $u^{(51)} - u$; the approximation is between 4 and 6 digits of accuracy away from ∂S. These graphs indicate that the accuracy deteriorates significantly close to the boundary, but improves considerably in the interior.

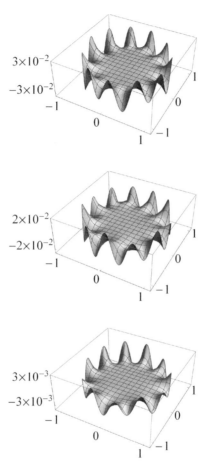

Fig. 5.2 Components of the error $u^{(51)} - u$

5.2.2 Computational Procedure

The step-by-step outline of the computational procedure used for solving the interior Robin problem starts with the boundary data function $\mathcal{R}$, whose components (in polar coordinates) are graphed in Fig. 5.3.

$\mathscr{R}$ is digitized on ∂S with a sufficient number of points to guarantee that the accuracy of all quadrature calculations significantly exceeds the expected magnitude of the error produced by our approximation method. This allows the method error to be isolated from any quadrature errors. For this example, 100 discrete points on ∂S prove adequate to reach our objective. Next, the functions $\varphi^{(i)}$ are constructed as described in (5.21) and (5.22), and are digitized as well. An orthonormalization method is then selected—in our case, MGS—to generate the digitized $\omega^{(i)}$ as described earlier.

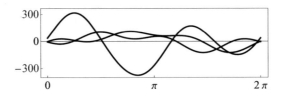

Fig. 5.3 Components of $\mathscr{R}$ in polar coordinates

The graphs of typical three-component vector functions $\varphi^{(i)}$ and $\omega^{(i)}$ are displayed in Figs. 5.4 and 5.5, respectively. The increasing number of oscillations in the latter eventually produces computational difficulties in the evaluation of the inner products occurring in the construction of additional orthonormal vector functions.

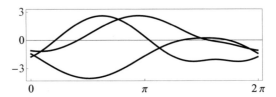

Fig. 5.4 Typical vector functions $\varphi^{(i)}$ in polar coordinates

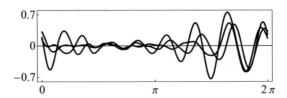

Fig. 5.5 Typical vector functions $\omega^{(i)}$ in polar coordinates

5.2.3 Computational Results

For the interior Robin problem, the approximation $\psi^{(n)}$ of $u|_{\partial S}$ is computed by means of the system of equations generated by (5.23), (5.24), (5.26), and the $x^{(k)}$.

In our example, $\psi^{(51)}$ was calculated with 16 points $x^{(k)}$ on ∂S_* (see Fig. 5.6). As Fig. 5.7 shows, this approximation is quite good when compared with the exact result for $u|_{\partial S}$ obtained from the known solution u.

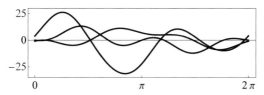

Fig. 5.6 Components of $\psi^{(51)}$ in polar coordinates

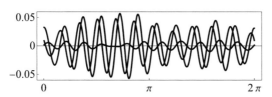

Fig. 5.7 Components of the error $\psi^{(51)} - u|_{\partial S}$ in polar coordinates

Equation (5.32) and $\psi^{(51)}$ generate the results displayed in Figs. 5.1 and 5.2. Finally, we can compare the computed approximation

$$\mathcal{R}^{(51)} = Tu + \sigma\psi^{(51)} \tag{5.35}$$

to the actual boundary condition $\mathcal{R}$. The first term on the right-hand side in (5.35) is constructed from the known exact solution $u(x)$, $x \in S^+$, as

$$Tu(x_p) = \begin{pmatrix} 6 - 8\cos\theta + 10\sin\theta + 4\cos(2\theta) + 3\cos(3\theta) + 6\sin(3\theta) \\ 10\cos\theta + 8\sin\theta + 4\sin(2\theta) - 6\cos(3\theta) - 9\sin(3\theta) \\ 12\cos\theta - 18\cos(2\theta) + 36\sin(2\theta) \end{pmatrix}.$$

The components of $\mathcal{R}^{(51)}$ are displayed in Fig. 5.8. As Fig. 5.9 shows, this approximation is in good agreement with the exact boundary data function $\mathcal{R}$.

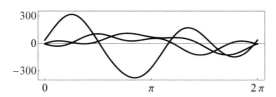

Fig. 5.8 Components of $\mathcal{R}^{(51)}$ in polar coordinates

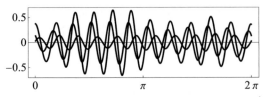

Fig. 5.9 Components of $\mathscr{R}^{(51)} - \mathscr{R}$ in polar coordinates

5.2.4 Error Analysis

The approximation $(u|_{\partial S})^{(n)} = \psi^{(n)}$, $n = 3N + 3$, is computed using N equally spaced points $x^{(k)}$ on ∂S_*. The behavior of the relative error

$$\frac{\|\psi^{(3N+3)} - u|_{\partial S}\|}{\|u|_{\partial S}\|}$$

as a function of N, illustrated by the logarithmic plot in Fig. 5.10, determines the efficiency and accuracy of the computational procedure.

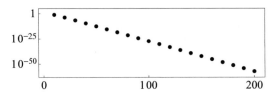

Fig. 5.10 The relative error $\|\psi^{(3N+3)} - u|_{\partial S}\|/\|u|_{\partial S}\|$ for $N = 5, 10, \ldots, 200$

This plot strongly suggests that the relative error improves exponentially as N increases. Fitting a linear curve to the logarithmic data produces the model

$$2.28505 - 0.296757 N.$$

A comparison of the logarithmic data and the linear model is shown in Fig. 5.11.

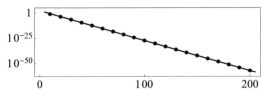

Fig. 5.11 The relative error $\|\psi^{(3N+3)} - u|_{\partial S}\|/\|u|_{\partial S}\|$ and the linear model for $N = 5, 10, \ldots, 200$

The two are seen to be in good agreement; consequently, the relative error may be modeled by

$$\frac{\|\psi^{(3N+3)} - u|_{\partial S}\|}{\|u|_{\partial S}\|} = 192.773 \times 10^{-0.296757N} = 192.773 \times 0.504944^N.$$

The exponentially decreasing error is optimal in the sense that its rate of improvement will eventually exceed that of any fixed-order method. Our model indicates that each additional point $x^{(k)}$ added on ∂S_* will reduce the relative error by about 50%. Clearly, these numbers are specific to our example and should not to be interpreted as universally valid.

5.3 Numerical Example 2

With the same plate parameters, auxiliary circle ∂S_*, and points $x^{(k)}$ as in Sect. 5.2, and with

$$\sigma(x_p) = \begin{pmatrix} 10 & -4 & -1+3\cos(2\theta) \\ -4 & 10 & 1-3\cos(2\theta) \\ -1+3\cos(2\theta) & 1-3\cos(2\theta) & 13+3\cos(2\theta) \end{pmatrix},$$

we choose the boundary condition, written in polar coordinates, in (5.1) to be

$$\mathcal{R}(x_p) = \frac{1}{2} \begin{pmatrix} -31 - 43\cos\theta + 9\sin\theta - 90\cos(2\theta) + 311\sin(2\theta) \\ \quad - 167\cos(3\theta) - 216\sin(3\theta) + 29\cos(4\theta) + 82\sin(4\theta) \\ \quad + 99\sin(5\theta) + 6\cos(6\theta) + 3\sin(6\theta) \\ -23 + 21\cos\theta + 125\sin\theta + 278\cos(2\theta) - 123\sin(2\theta) \\ \quad - 13\cos(3\theta) + 276\sin(3\theta) - 89\cos(4\theta) - 22\sin(4\theta) \\ \quad - 99\sin(5\theta) - 6\cos(6\theta) - 3\sin(6\theta) \\ 100 - 171\cos\theta + 159\sin\theta + 149\cos(2\theta) - 14\sin(2\theta) \\ \quad - 13\cos(3\theta) + 1074\sin(3\theta) + 16\cos(4\theta) + 59\sin(4\theta) \\ \quad - 12\cos(5\theta) + 63\sin(5\theta) + 15\cos(6\theta) + 12\sin(6\theta) \end{pmatrix}.$$

The exact solution (in S^+) for this function is

$$u(x)$$
$$= \begin{pmatrix} -2x_1 - 3x_1^2 - x_2^2 - 8x_1^3 - 12x_1^2 x_2 + 24x_1 x_2^2 + 4x_2^3 + 36x_1^3 x_2 + 12x_1 x_2^3 \\ -2 + 10x_2 - 2x_1 x_2 - 4x_1^3 + 24x_1^2 x_2 + 12x_1 x_2^2 - 8x_2^3 + 9x_1^4 + 18x_1^2 x_2^2 - 15x_2^4 \\ 7 - 6x_1 + 2x_2 + x_1^2 - 5x_2^2 + x_1^3 + 108x_1^2 x_2 + x_1 x_2^2 - 36x_2^3 + 2x_1^4 + 4x_1^3 x_2 \\ \quad - 12x_1^2 x_2^2 - 4x_1 x_2^3 + 2x_2^4 - 9x_1^4 x_2 - 6x_1^2 x_2^3 + 3x_2^5 \end{pmatrix}.$$

Applying the procedure described in Remark 1.11, we find that the solution contains the rigid displacement $f = 6f^{(1)} - 2f^{(2)} - 7f^{(3)}$.

The details mentioned in Remarks 3.9–3.12 apply here as well.

We approximate the solution of problem (5.1) in this example by means of the MGS method.

5.3.1 Graphical Illustrations

The graphs of the components of $u^{(51)}$ computed from $\psi^{(51)}$ (that is, with 16 points $x^{(k)}$ on ∂S_*) for

$$0 \le r \le 0.97, \quad 0 \le \theta < 2\pi,$$

together with those of the components of $\psi^{(51)} = (u|_{\partial S})^{(51)}$, are shown in Fig. 5.12.

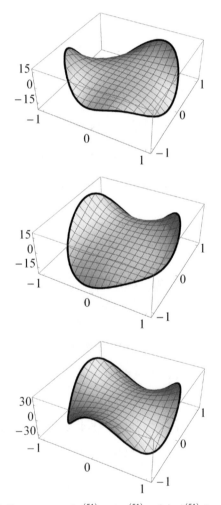

Fig. 5.12 Components of $u^{(51)}$ and $\psi^{(51)} = (u|_{\partial S})^{(51)}$ (heavy lines)

The reason for the restriction $r \leq 0.97$ is the increasing influence of the singularities of $D(x,y)$ and $P(x,y)$ for $x \in S^+$ very close to $y \in \partial S$. This drawback can be mitigated by increasing the floating-point accuracy in the vicinity of ∂S, but it is never completely eliminated.

Figure 5.13 contains the graphs of the components of the error $u^{(51)} - u$; the approximation is between 4 and 6 digits of accuracy away from ∂S. These graphs indicate that the accuracy deteriorates significantly close to the boundary, but improves considerably in the interior.

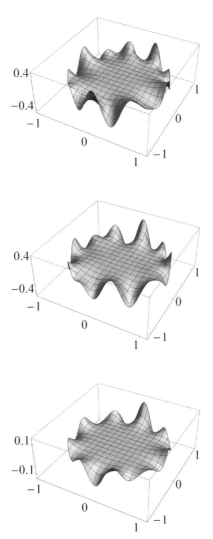

Fig. 5.13 Components of the error $u^{(51)} - u$

5.3.2 Computational Procedure

The step-by-step outline of the computational procedure used for solving the interior
Robin problem starts with the boundary data function $\mathscr{R}$, whose components (in
polar coordinates) are graphed in Fig. 5.14.

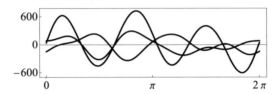

Fig. 5.14 Components of $\mathscr{R}$ in polar coordinates

$\mathscr{R}$ is now digitized on ∂S and the functions $\varphi^{(i)}$ are constructed and digitized
as described in Sect. 5.2.2. An orthonormalization method is then selected—in our
case, MGS—to generate the digitized $\omega^{(i)}$ as described earlier.

The graphs of typical three-component vector functions $\varphi^{(i)}$ and $\omega^{(i)}$ are displayed in Figs. 5.15 and 5.16, respectively. The increasing number of oscillations
in the latter eventually produces computational difficulties in the evaluation of the
inner products occurring in the construction of additional orthonormal vector functions.

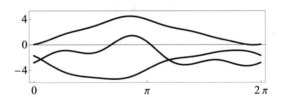

Fig. 5.15 Typical vector functions $\varphi^{(i)}$ in polar coordinates

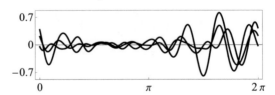

Fig. 5.16 Typical vector functions $\omega^{(i)}$ in polar coordinates

5.3.3 Computational Results

For the interior Robin problem, the approximation $\psi^{(n)}$ of $u|_{\partial S}$ is computed by means of the system of equations generated by (5.23), (5.24), (5.26), and the $x^{(k)}$.

In our example, $\psi^{(51)}$ was constructed with 16 points $x^{(k)}$ on ∂S_* (see Fig. 5.17). As Fig. 5.18 shows, this approximation is quite good when compared with the exact result for $u|_{\partial S}$ obtained from the known solution u.

Equation (5.32) and $\psi^{(51)}$ generate the results displayed in Figs. 5.12 and 5.13. Finally, we can compare the computed approximation

$$\mathcal{R}^{(51)} = Tu + \sigma\psi^{(51)} \tag{5.36}$$

to the actual boundary condition $\mathcal{R}$. The first term on the right-hand side in (5.36)

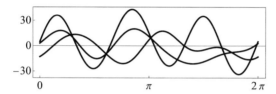

Fig. 5.17 Components of $\psi^{(51)}$ in polar coordinates

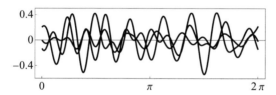

Fig. 5.18 Components of the error $\psi^{(51)} - u|_{\partial S}$ in polar coordinates

is calculated from the known exact solution u as

$$Tu(x_p) = \begin{pmatrix} -3 + \cos\theta - 2\cos(2\theta) + 30\sin(2\theta) - 12\cos(3\theta) \\ -6\sin(3\theta) + 12\sin(4\theta) \\ 7\sin\theta + 30\cos(2\theta) - 2\sin(2\theta) - 6\cos(3\theta) \\ +12\sin(3\theta) - 12\cos(4\theta) \\ -6\cos\theta - 108\sin(3\theta) \end{pmatrix}.$$

The components of $\mathcal{R}^{(51)}$ are displayed in Fig. 5.19. As Fig. 5.20 shows, this approximation is in good agreement with the exact boundary data function $\mathcal{R}$.

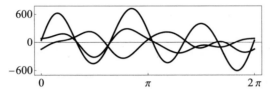

Fig. 5.19 Components of $\mathscr{R}^{(51)}$ in polar coordinates

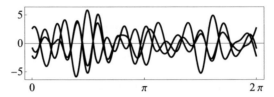

Fig. 5.20 Components of $\mathscr{R}^{(51)} - \mathscr{R}$ in polar coordinates

5.3.4 Error Analysis

The approximation $(u|_{\partial S})^{(n)} = \psi^{(n)}$, $n = 3N + 3$, is computed with N equally spaced points $x^{(k)}$ on ∂S_*. The behavior of the relative error

$$\frac{\|\psi^{(3N+3)} - u|_{\partial S}\|}{\|u|_{\partial S}\|}$$

as a function of N, illustrated by the logarithmic plot in Fig. 5.21, determines the efficiency and accuracy of the computational procedure.

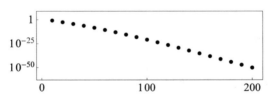

Fig. 5.21 The relative error $\|\psi^{(3N+3)} - u|_{\partial S}\| / \|u|_{\partial S}\|$ for $N = 5, 10, \ldots, 200$

This plot strongly suggests that the relative error improves exponentially as N increases. Fitting a linear curve to the logarithmic data (with the first 5 data points removed) produces the model

$$8.59682 - 0.292962\,N.$$

A comparison of the logarithmic data and the linear model is shown in Fig. 5.22.

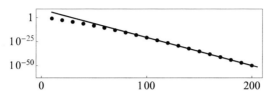

Fig. 5.22 The relative error $\|\psi^{(3N+3)} - u|_{\partial S}\|/\|u|_{\partial S}\|$ and the linear model for $N = 5, 10, \ldots, 200$

The two are seen to be in good agreement for $N > 50$; consequently, the relative error may be modeled by

$$\frac{\|\psi^{(3N+3)} - u|_{\partial S}\|}{\|u|_{\partial S}\|} = 3.95199 \times 10^8 \times 10^{-0.292962N}$$

$$= 3.95199 \times 10^8 \times 0.509376^N.$$

The exponentially decreasing error is optimal in the sense that its rate of improvement will eventually exceed that of any fixed-order method. Our model indicates that each additional point $x^{(k)}$ added on ∂S_* will reduce the relative error by about 50%. Clearly, these numbers are specific to our example and should not to be interpreted as universally valid.

5.4 Numerical Example 3

With the same plate parameters, auxiliary circle ∂S_*, and points $x^{(k)}$ as in Sect. 5.2, and with

$$\sigma(x) = \begin{pmatrix} 11 & 2 & 1 \\ 2 & 8 & -2 \\ 1 & -2 & 11 \end{pmatrix},$$

we choose the boundary condition, written in polar coordinates, in (5.1) to be

$$\mathscr{R}(x_p) = \begin{pmatrix} -24 + 41\cos\theta - 100\sin\theta + 75\cos(2\theta) + 32\sin(2\theta) \\ \quad + 39\cos(3\theta) + 28\sin(3\theta) + 11\cos(4\theta) + 32\sin(4\theta) \\ -64\cos\theta - 59\sin\theta + 4\cos(2\theta) - 46\sin(2\theta) - 18\cos(3\theta) \\ \quad - 26\sin(3\theta) - 22\cos(4\theta) + 16\sin(4\theta) \\ -24 + 42\cos\theta + 223\cos(2\theta) + 54\sin(2\theta) + 159\cos(3\theta) \\ \quad + 332\sin(3\theta) + 5\cos(4\theta) \end{pmatrix}.$$

The exact solution u of the problem is not known in this case.
The details mentioned in Remarks 3.9–3.12 apply here as well.

We approximate the solution of problem (5.1) in this example by means of the MRR method.

5.4.1 Graphical Illustrations

The graphs of the components of $u^{(51)}$ computed from $\psi^{(51)}$ (that is, with 16 points $x^{(k)}$ on ∂S_*) for

$$0 \le r \le 0.97, \quad 0 \le \theta < 2\pi,$$

together with those of the components of $\psi^{(51)} = (u|_{\partial S})^{(51)}$, are shown in Fig. 5.23.

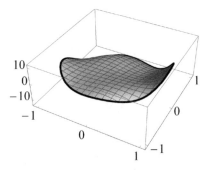

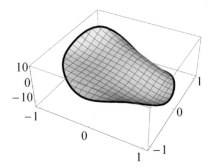

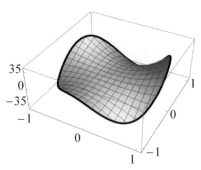

Fig. 5.23 Components of $u^{(51)}$ and $\psi^{(51)} = (u|_{\partial S})^{(51)}$ (heavy lines)

5.4.2 Computational Procedure

The step-by-step outline of the computational procedure used for solving the interior Robin problem starts with the boundary data function $\mathscr{R}$, whose components (in polar coordinates) are graphed in Fig. 5.24.

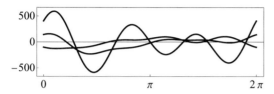

Fig. 5.24 Components of $\mathscr{R}$ in polar coordinates

$\mathscr{R}$ is now digitized on ∂S and the functions $\varphi^{(i)}$ are constructed and digitized as described in Sect. 5.2.2.

5.4.3 Computational Results

For the interior Robin problem, the approximation $\psi^{(n)}$ of $u|_{\partial S}$ is computed by means of the system of equations generated by (5.23), (5.24), (5.26), and the $x^{(k)}$. In our example, $\psi^{(51)}$ was calculated with 16 points $x^{(k)}$ on ∂S_* (see Fig. 5.25). This approximation was then combined with $u^{(51)}$ to generate the graphs displayed in Fig. 5.23, which show good agreement between $u^{(51)}$ and $\psi^{(51)}$. This is an indirect validation of our method.

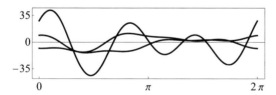

Fig. 5.25 Components of $\psi^{(51)}$ in polar coordinates

5.4.4 Error Analysis

The approximation $(u|_{\partial S})^{(n)} = \psi^{(n)}$, $n = 3N+3$, is computed with N equally spaced points $x^{(k)}$ on ∂S_*. The behavior of the relative error

$$\frac{\|\psi^{(3N+3)} - u|_{\partial S}\|}{\|u|_{\partial S}\|}$$

as a function of N determines the efficiency and accuracy of the computational procedure. However, since the exact expression of $u|_{\partial S}$ is not known in this example, an accurate error analysis cannot be performed.

As a substitute, we propose instead an indirect route to obtaining a measure of partial validation of our calculations. We use $\psi^{(51)} = (u|_{\partial S})^{(51)}$ as the boundary condition for an interior Dirichlet problem, and use the approximate solution of the latter to compute (by means of the procedure laid out in Chap. 3) an approximation $\Gamma^{(51)}$ of Tu (on ∂S), which, if our method is sound, should lead to a close fit to $\mathscr{R}$. Figure 5.26 shows the components of $\Gamma^{(51)}$.

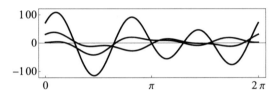

Fig. 5.26 Components of $\Gamma^{(51)}$ in polar coordinates

Combining $\psi^{(51)}$ and $\Gamma^{(51)}$ yields the approximation of the Robin boundary condition

$$\mathscr{R}^{(51)} = \Gamma^{(51)} + \sigma\psi^{(51)}.$$

The components of $\mathscr{R}^{(51)}$ and those of the difference $\mathscr{R}^{(51)} - \mathscr{R}$ are shown in Figs. 5.27 and 5.28, respectively.

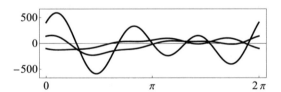

Fig. 5.27 Components of $\mathscr{R}^{(51)}$ in polar coordinates

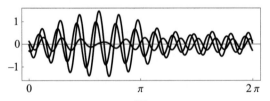

Fig. 5.28 Components of $\mathscr{R}^{(51)} - \mathscr{R}$ in polar coordinates

Figure 5.28 indicates that the total combined error in this verification exercise is consistent with the error expected when we use MRR to compute $\mathscr{R}^{(51)}$. Repeating our numerical experiment with 200 points $x^{(k)}$ instead of 16 on ∂S_*, we obtained the much smaller relative error

$$\frac{\|\mathscr{R}^{(603)} - \mathscr{R}\|}{\|\mathscr{R}\|} = 11.7317 \times 10^{-57},$$

which confirms the robust nature of the generalized Fourier series method of approximation presented here.

5.5 Numerical Example 4

With the same plate parameters, auxiliary circle ∂S_*, and points $x^{(k)}$ as in Sect. 5.2, and with

$$\sigma(x_p) = \begin{pmatrix} 10 & -4 & -1+3\cos(2\theta) \\ -4 & 10 & 1-3\cos(2\theta) \\ -1+3\cos(2\theta) & 1-3\cos(2\theta) & 13+3\cos(2\theta) \end{pmatrix},$$

we choose the boundary condition, written in polar coordinates, in (5.1) to be

$$\mathscr{R}(x_p) = \frac{1}{2} \begin{pmatrix} 95 + 81\cos\theta - 36\sin\theta - 53\cos(2\theta) + 53\sin(2\theta) \\ \quad + 111\cos(3\theta) + 136\sin(3\theta) + 83\cos(4\theta) + 2\sin(4\theta) \\ \quad - 3\cos(6\theta) - 3\sin(6\theta) \\ -41 - 15\cos\theta - 30\sin\theta + 5\cos(2\theta) - 37\sin(2\theta) \\ \quad + 63\cos(3\theta) - 82\sin(3\theta) - 83\cos(4\theta) - 2\sin(4\theta) \\ \quad + 3\cos(6\theta) + 3\sin(6\theta) \\ 136 - 261\cos\theta + 300\sin\theta + 817\cos(2\theta) - 9\sin(2\theta) \\ \quad - 24\cos(3\theta) + 35\sin(3\theta) + 58\cos(4\theta) - 17\sin(4\theta) \\ \quad + 9\cos(5\theta) + 15\sin(5\theta) - 3\cos(6\theta) - 3\sin(6\theta) \end{pmatrix}.$$

The exact solution u of the problem is not known in this case.

The details mentioned in Remarks 3.9–3.12 apply here as well.

We approximate the solution of problem (5.1) in this example by means of the MRR method.

5.5.1 Graphical Illustrations

The graphs of the components of $u^{(51)}$ computed from $\psi^{(51)}$ (that is, with 16 points $x^{(k)}$ on ∂S_*) for
$$0 \le r \le 0.97, \quad 0 \le \theta < 2\pi,$$
together with those of the components of $\psi^{(51)} = (u|_{\partial S})^{(51)}$, are shown in Fig. 5.29.

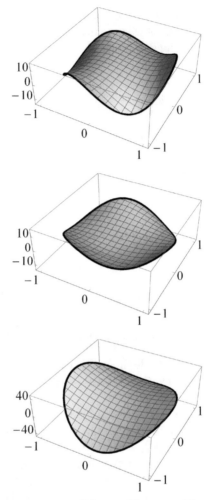

Fig. 5.29 Components of $u^{(51)}$ and $\psi^{(51)} = (u|_{\partial S})^{(51)}$ (heavy lines)

5.5.2 Computational Procedure

The step-by-step outline of the computational procedure used for solving the interior Robin problem starts with the boundary data function $\mathscr{R}$, whose components (in polar coordinates) are graphed in Fig. 5.30.

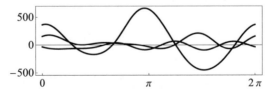

Fig. 5.30 Components of $\mathscr{R}$ in polar coordinates

$\mathcal{R}$ is now digitized on ∂S and the functions $\varphi^{(i)}$ are constructed and digitized as described in Sect. 5.2.2.

5.5.3 Computational Results

For the interior Robin problem, the approximation $\psi^{(n)}$ of $u|_{\partial S}$ is constructed by means of the system of equations generated by (5.23), (5.24), (5.26), and the $x^{(k)}$. In our example, $\psi^{(51)}$ was calculated with 16 points $x^{(k)}$ on ∂S_* (see Fig. 5.31). This approximation was then combined with $u^{(51)}$ to generate the graphs displayed in Fig. 5.29, which show good agreement between $u^{(51)}$ and $\psi^{(51)}$. This is an indirect validation of our method.

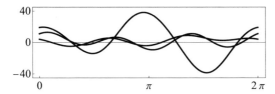

Fig. 5.31 Components of $\psi^{(51)}$ in polar coordinates

5.5.4 Error Analysis

The approximation $(u|_{\partial S})^{(n)} = \psi^{(n)}, n = 3N+3$, is computed with N equally spaced points $x^{(k)}$ on ∂S_*. The behavior of the relative error

$$\frac{\|\psi^{(3N+3)} - u|_{\partial S}\|}{\|u|_{\partial S}\|}$$

as a function of N determines the efficiency and accuracy of the computational procedure. However, since the exact expression of $u|_{\partial S}$ is not known in this example, an accurate error analysis cannot be performed.

As a substitute, we propose instead an indirect route to obtaining a measure of partial validation of our calculations. We use $\psi^{(51)} = (u|_{\partial S})^{(51)}$ as the boundary condition for an interior Dirichlet problem, and use the approximate solution of the latter to compute (by means of the procedure laid out in Chap. 3) an approximation $\Gamma^{(51)}$ of Tu (on ∂S), which, if our method is sound, should lead to a close fit to $\mathcal{R}$. Figure 5.32 shows the components of $\Gamma^{(51)}$.

Combining $\psi^{(51)}$ and $\Gamma^{(51)}$ yields the approximation of the Robin boundary condition

$$\mathcal{R}^{(51)} = \Gamma^{(51)} + \sigma\psi^{(51)}.$$

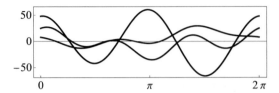

Fig. 5.32 Components of $\Gamma^{(51)}$ in polar coordinates

The components of $\mathscr{R}^{(51)}$ and those of the difference $\mathscr{R}^{(51)} - \mathscr{R}$ are shown in Figs. 5.33 and 5.34, respectively.

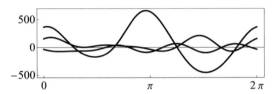

Fig. 5.33 Components of $\mathscr{R}^{(51)}$ in polar coordinates

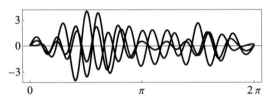

Fig. 5.34 Components of $\mathscr{R}^{(51)} - \mathscr{R}$ in polar coordinates

Figure 5.34 indicates that the total combined error in this verification exercise is consistent with the error expected when we use MRR to compute $\mathscr{R}^{(51)}$. Repeating our numerical experiment with 200 points $x^{(k)}$ instead of 16 on ∂S_*, we obtained the much smaller relative error

$$\frac{\|\mathscr{R}^{(603)} - \mathscr{R}\|}{\|\mathscr{R}\|} = 8.04755 \times 10^{-50},$$

which confirms the robust nature of the generalized Fourier series method of approximation presented here.

Chapter 6
Exterior Dirichlet Problem

6.1 Computational Algorithm

The exterior Dirichlet problem consists in finding u such that

$$Au(x) = 0, \quad x \in S^-,$$
$$u(x) = \mathscr{D}(x), \quad x \in \partial S, \qquad (6.1)$$
$$u \in \mathscr{A}^*,$$

where the vector function $\mathscr{D} \in C^{(1,\alpha)}(\partial S)$, $\alpha \in (0,1)$, is prescribed.

The representation formulas (1.22) hold for a solution of the system that belongs to the asymptotic class $\mathscr{A}$. Writing

$$u = u^{\mathscr{A}} + Fa, \qquad (6.2)$$

where $u^{\mathscr{A}} \in \mathscr{A}$ and Fa is the rigid displacement component of u, we see that, since $A(Fa) = 0$ in $\mathbb{R}^2$ and $T(Fa) = 0$ on ∂S, the vector function $u^{\mathscr{A}}$ is a solution of the exterior Dirichlet problem

$$Au^{\mathscr{A}}(x) = 0, \quad x \in S^-,$$
$$u^{\mathscr{A}}(x) = (\mathscr{D} - Fa)(x), \quad x \in \partial S,$$
$$u^{\mathscr{A}} \in \mathscr{A};$$

therefore, by (1.22),

$$u^{\mathscr{A}}(x) = -\int_{\partial S} D(x,y)Tu^{\mathscr{A}}(y)\,ds(y) + \int_{\partial S} P(x,y)(\mathscr{D} - Fa)(y)\,ds(y), \quad x \in S^-,$$

$$0 = -\int_{\partial S} D(x,y)Tu^{\mathscr{A}}(y)\,ds(y) + \int_{\partial S} P(x,y)(\mathscr{D} - Fa)(y)\,ds(y), \quad x \in S^+.$$

© Springer Nature Switzerland AG 2020
C. Constanda, D. Doty, *The Generalized Fourier Series Method*, Developments in Mathematics 65, https://doi.org/10.1007/978-3-030-55849-9_6

In view of Remark 1.36(ii), the fact that

$$\langle Tu, f^{(i)} \rangle = \langle Tu^{\mathscr{A}}, f^{(i)} \rangle = \langle \psi, f^{(i)} \rangle = 0, \quad i = 1, 2, 3,$$

and in accordance with Remark 1.24, the above formulas can be rewritten as

$$u^{\mathscr{A}}(x) = - \int_{\partial S} (D(x,y))^{\mathscr{A}} Tu^{\mathscr{A}}(y) \, ds(y)$$

$$+ \int_{\partial S} P(x,y) \mathscr{D}(y) \, ds(y), \quad x \in S^-, \tag{6.3}$$

$$0 = - \int_{\partial S} (D(x,y))^{\mathscr{A}} Tu^{\mathscr{A}}(y) \, ds(y)$$

$$+ \int_{\partial S} P(x,y) \mathscr{D}(y) \, ds(y) + (Fa)(x), \quad x \in S^+. \tag{6.4}$$

Let ∂S_* be a simple, closed, C^2–curve lying strictly inside S^+, and let $\{x^{(k)}\}_{k=1}^{\infty}$ be a set of points densely distributed on ∂S_*. Adopting the notational substitutions

$$(D(y,x))^{\mathscr{A}} \big|_{x \to x^{(k)}, \, y \to x} = (D(x, x^{(k)}))^{\mathscr{A}},$$
$$(D(x,y))^{\mathscr{A}} \big|_{x \to x^{(k)}, \, y \to x} = (D(x^{(k)}, x))^{\mathscr{A}}, \tag{6.5}$$

we consider the sequence of vector functions on ∂S defined by

$$\mathscr{G} = \{f^{(i)}, \, i = 1,2,3\} \cup \{\varphi^{(jk)}, \, j = 1,2,3, \, k = 1,2,\ldots\},$$

where

$$\varphi^{(jk)}(x) = \left((D(x, x^{(k)}))^{\mathscr{A}} \right)^{(j)}, \quad x \in \partial S, \, j = 1,2,3, \, k = 1,2,\ldots. \tag{6.6}$$

6.1 Remark. The vector functions $\varphi^{(jk)}$ have an alternative expression, more suitable for use in certain cases. Transposing the sides of the first equality in (1.20), we see that

$$(D(y,x))^{\mathscr{A}} = \left((D(x,y))^{\mathscr{A}} \right)^{\mathsf{T}}, \quad x \neq y,$$

from which, by Remark 3.2,

$$\left((D(y,x))^{\mathscr{A}} \right)^{(j)} = \left(\left((D(x,y))^{\mathscr{A}} \right)^{\mathsf{T}} \right)^{(j)} = \left(\left((D(x,y))^{\mathscr{A}} \right)_{(j)} \right)^{\mathsf{T}}, \quad x \neq y.$$

Taking (6.6) and (6.5) into account, we now deduce that

$$\varphi^{(jk)}(x) = \left((D(x, x^{(k)}))^{\mathscr{A}} \right)^{(j)} = \left(\left((D(x^{(k)}, x))^{\mathscr{A}} \right)_{(j)} \right)^{\mathsf{T}}. \tag{6.7}$$

6.2 Theorem. *The set $\mathscr{G}$ is linearly independent on ∂S and complete in $L^2(\partial S)$.*

Proof. Assuming the opposite, suppose that there are a positive integer N and real numbers c_i and c_{jk}, $i,j = 1,2,3$, $k = 1,2,\ldots,N$, not all zero, such that

$$\sum_{i=1}^{3} c_i f^{(i)}(x) + \sum_{j=1}^{3}\sum_{k=1}^{N} c_{jk}\varphi^{(jk)}(x) = 0, \quad x \in \partial S. \tag{6.8}$$

We set

$$g(x) = \sum_{i=1}^{3} c_i f^{(i)}(x) + \sum_{j=1}^{3}\sum_{k=1}^{N} c_{jk}\varphi^{(jk)}(x)$$

$$= \sum_{i=1}^{3} c_i f^{(i)}(x) + \sum_{j=1}^{3}\sum_{k=1}^{N} c_{jk}\big((D(x,x^{(k)}))^{\mathscr{A}}\big)^{(j)}$$

and see that, by (6.6), (6.8), and Theorem 1.9 applied to $D^{\mathscr{A}}$,

$$Ag = 0 \quad \text{in } S^{-},$$
$$g = 0 \quad \text{on } \partial S,$$
$$g \in \mathscr{A}^{*};$$

that is, g is the (unique) regular solution of the homogeneous Dirichlet problem in S^{-}. According to Theorem 1.29(iv),

$$g = 0 \quad \text{in } \bar{S}^{-}.$$

By Corollary 3.12, p.135 in [14],

$$c_i = 0, \quad i = 1,2,3,$$

and

$$\sum_{j=1}^{3}\sum_{k=1}^{N} c_{jk}\varphi^{(jk)}(x) = 0, \quad x \in S^{-}. \tag{6.9}$$

But, according to the asymptotic expansions in Sect. 1.2.1,

$$\left|\big(((D(x,x^{(k)}))^{\mathscr{A}})^{(j)}\big)_3\right| = \left|\big((D(x,x^{(k)}))^{\mathscr{A}}\big)_{3j}\right| \to \infty \text{ as } |x| \to \infty$$

since it contains a $\ln|x|$ term for every $j = 1,2,3$ and $k = 1,2,\ldots,N$, which contradicts (6.9), so all $c_{jk} = 0$, thus proving the linear independence of $\mathscr{G}$.

For the second part of the proof, suppose that there is $q \in L^2(\partial S)$ such that

$$\langle q, f^{(i)}\rangle = \int_{\partial S} (f^{(i)})^{\mathsf{T}} q\, ds = 0, \quad i = 1,2,3, \tag{6.10}$$

$$\langle q, \varphi^{(jk)}\rangle = \int_{\partial S} (\varphi^{(jk)})^{\mathsf{T}} q\, ds = 0, \quad j = 1,2,3,\ k = 1,2,\ldots. \tag{6.11}$$

By (6.7),

$$\left(\varphi^{(jk)}\right)^{\mathsf{T}} = \left(\left(D(x^{(k)}, x)\right)^{\mathscr{A}}\right)_{(j)},$$

so equality (6.11) is equivalent to

$$\int_{\partial S} \left(D(x^{(k)}, y)\right)^{\mathscr{A}} q(y)\, ds(y) = 0, \quad k = 1, 2, \dots. \tag{6.12}$$

By Theorem 4.18(i), p. 99 in [15], the single-layer potential with an L^2-density

$$(Vq)(x) = \int_{\partial S} D(x, y) q(y)\, ds(y), \quad x \in S^+ \cup S^-,$$

is continuous on ∂S_*. Since, in view of (6.10),

$$(Vq)(x) = (Vq)^{\mathscr{A}}(x) = \int_{\partial S} \left(D(x, y)\right)^{\mathscr{A}} q(y)\, ds(y), \tag{6.13}$$

and since the points $x^{(k)}$, $k = 1, 2, \dots$, are densely distributed on ∂S_*, from (6.12) we deduce that $Vq = 0$ on ∂S_*. Thus, (6.10) and Theorem 4.18(ii),(iii), p. 99 in [15] yield

$$A(Vq) = 0 \quad \text{in } S_*^+,$$
$$Vq = 0 \quad \text{on } \partial S_*,$$

which shows that Vq is the (unique) solution of the homogeneous interior Dirichlet problem in $\bar{S}_*^+$; hence,

$$Vq = 0 \quad \text{in } \bar{S}_*^+.$$

Since the single-layer potential Vq is analytic in $S^+ \cup S^-$, it follows that

$$Vq = 0 \quad \text{in } \bar{S}^+. \tag{6.14}$$

Consequently, $\left(T(Vq)\right)^+ = 0$, so from Theorem 4.21, p.100 in [15] it follows that

$$\tfrac{1}{2} q(x) + \int_{\partial S} T(\partial_x) D(x, y) q(y)\, ds(y) = 0 \quad \text{for a.a. } x \in \partial S,$$

with the integral above understood as principal value. According to Theorem 6.12, p. 143 in [15], $q \in C^{0,\alpha}(\partial S)$ for any $\alpha \in (0, 1)$. Hence, by Theorem 1.9(i) in [15], Vq is continuous in $\mathbb{R}^2$ and, in view of (6.14) and (6.13), we have

$$A(Vq) = 0 \quad \text{in } S^-,$$
$$Vq = 0 \quad \text{on } \partial S,$$
$$Vq \in \mathscr{A}.$$

This shows that Vq is the (unique) regular solution of the homogeneous Dirichlet problem in $\bar{S}^-$. Consequently, according to Theorem 1.29(iv),

$$Vq = 0 \quad \text{in } \bar{S}^-,$$

which implies that

$$(T(Vq))^- = 0 \quad \text{on } \partial S.$$

By Theorem 3.22, p. 140 in [14], the null space of the boundary integral operator defined by the above equality consists of the zero vector alone, so $q = 0$.

Given that $L^2(\partial S)$ is a Hilbert space, Theorem 3.1 now implies that the set $\mathcal{G}$ is complete in $L^2(\partial S)$. □

We order the elements of $\mathcal{G}$ as the sequence

$$f^{(1)}, f^{(2)}, f^{(3)}, \varphi^{(11)}, \varphi^{(21)}, \varphi^{(31)}, \varphi^{(12)}, \varphi^{(22)}, \varphi^{(32)}, \ldots$$

and re-index them, noticing that for each $i = 4, 5, \ldots$, there is a unique pair $\{j, k\}$, $j = 1, 2, 3$, $k = 1, 2, \ldots$, such that $i = j + 3k$. Then

$$\mathcal{G} = \{\varphi^{(1)}, \varphi^{(2)}, \varphi^{(3)}, \varphi^{(4)}, \varphi^{(5)}, \varphi^{(6)}, \varphi^{(7)}, \varphi^{(8)}, \varphi^{(9)}, \ldots\}, \tag{6.15}$$

where

$$\varphi^{(i)} = f^{(i)}, \quad i = 1, 2, 3, \tag{6.16}$$

and

$$\varphi^{(i)} = \varphi^{(jk)}, \quad j = 1, 2, 3, \ k = 1, 2, \ldots, \ i = j + 3k = 1, 2, \ldots. \tag{6.17}$$

6.1.1 Row Reduction Method

Our intention is to construct a nonsingular linear system of finitely many algebraic equations to approximate the solution u of (6.1).

Rewriting the representation formulas (6.3) and (6.4) as

$$u^{\mathscr{A}}(x) = -\langle D^{\mathscr{A}}(x, \cdot), Tu^{\mathscr{A}} \rangle + \langle P(x, \cdot), \mathscr{D} \rangle, \quad x \in S^-, \tag{6.18}$$

$$\langle D^{\mathscr{A}}(x, \cdot), Tu^{\mathscr{A}} \rangle = \langle P(x, \cdot), \mathscr{D} \rangle + (Fa)(x), \quad x \in S^+, \tag{6.19}$$

we would normally use (6.19) to approximate $Tu^{\mathscr{A}}$ (on ∂S), and then approximate $u^{\mathscr{A}}$ in S^- by means of (6.18). However, this operation cannot be performed since (6.19) contains the rigid displacement term

$$Fa = \sum_{i=1}^{3} a_i f^{(i)},$$

which may not be known a priori and needs to be determined as part of the computation process.

We set

$$\psi = Tu = Tu^{\mathscr{A}}.$$

Since the $x^{(k)}$ are points in S^+, from (6.4) it follows that

$$\int_{\partial S} \left(D(x^{(k)}, x) \right)^{\mathscr{A}} \psi(x)\, ds(x) = \int_{\partial S} P(x^{(k)}, x) \mathscr{D}(x)\, ds(x) + (Fa)(x^{(k)}),$$

which, by (6.7) and (6.17), leads to

$$\langle \varphi^{(i)}, \psi \rangle = \langle \varphi^{(jk)}, \psi \rangle = \int_{\partial S} \left(\left(D(x^{(k)}, x) \right)^{\mathscr{A}} \right)_{(j)} \psi(x)\, ds(x)$$

$$= \int_{\partial S} \left(P(x^{(k)}, x) \right)_{(j)} \mathscr{D}(x)\, ds(x) + \sum_{i=1}^{3} a_l f_j^{(l)}(x^{(k)}),$$

$$j = 1, 2, 3, \quad k = 1, 2, \ldots, \quad i = j + 3k = 4, 5, \ldots,$$

or, what is the same,

$$\langle \varphi^{(i)}, \psi \rangle = \left\langle \left(\left(\left(D(x^{(k)}, \cdot) \right)^{\mathscr{A}} \right)_{(j)} \right)^{\mathsf{T}}, \psi \right\rangle$$

$$= \left\langle \left(\left(P(x^{(k)}, \cdot) \right)_{(j)} \right)^{\mathsf{T}}, \mathscr{D} \right\rangle + \sum_{l=1}^{3} a_l f_j^{(l)}(x^{(k)}). \qquad (6.20)$$

In view of Theorem 6.2, there is a unique expansion

$$\psi = \sum_{h=1}^{\infty} c_h \varphi^{(h)}, \qquad (6.21)$$

so we can seek an approximation of ψ of the form

$$\psi^{(n)} = (Tu)^{(n)} = (Tu^{\mathscr{A}})^{(n)} = \sum_{h=1}^{n} c_h \varphi^{(h)}.$$

Replacing ψ by $\psi^{(n)}$ in (6.20), for $i = j + 3k = 4, 5, \ldots, n$ we arrive at the approximate equality

$$\sum_{h=1}^{n} c_h \langle \varphi^{(i)}, \varphi^{(h)} \rangle = \left\langle \left(\left(P(x^{(k)}, \cdot) \right)_{(j)} \right)^{\mathsf{T}}, \mathscr{D} \right\rangle + \sum_{l=1}^{3} a_l f_j^{(l)}(x^{(k)}). \qquad (6.22)$$

This is a linear system of $n - 3$ equations in $n + 3$ unknowns c_h and a_l. To have a consistent system with a unique solution, we need to augment (6.22) with six additional equations.

Let $\bar{x}$ be a point in S^+, distinct from the $x^{(k)}$ and the origin. Applying the same argument to (6.19) with $x = \bar{x}$, we obtain the additional approximate equality

$$\sum_{h=1}^{n} c_h \left\langle \left(\left((D(\bar{x}, \cdot))^{\mathscr{A}} \right)_{(i)} \right)^{\mathsf{T}}, \varphi^{(h)} \right\rangle$$
$$= \left\langle \left((P(\bar{x}, x))_{(i)} \right)^{\mathsf{T}}, \mathscr{D} \right\rangle + \sum_{l=1}^{3} a_l f_i^{(l)}(\bar{x}), \quad i = 1, 2, 3. \tag{6.23}$$

At the same time, since $\psi = Tu$ is the Neumann data of the problem, we have

$$\langle \psi, f^{(i)} \rangle = 0, \quad i = 1, 2, 3, \tag{6.24}$$

so it is natural to ask the approximation $\psi^{(n)}$ to satisfy the same condition; in other words,

$$\sum_{h=1}^{n} \langle \varphi^{(h)}, f^{(i)} \rangle c_h = 0, \quad i = 1, 2, 3. \tag{6.25}$$

Equations (6.25) and (6.23) are adjoined to system (6.22) to make it complete for the computation of the n coefficients c_h and the three coefficients a_l; in other words, for fully determining $\psi^{(n)}$ and the approximation

$$(Fa)^{(n)} = Fa^{(n)}$$

of the rigid displacement Fa.

6.3 Remark. For symmetry and ease of computation, in the construction of the approximation $\psi^{(n)}$ we use the subsequence with $n = 3N + 3$, where N is the number of points $x^{(k)}$ chosen on ∂S_*. This choice feeds into the computation, for each $k = 1, 2, \ldots, N$, all three columns/rows of $D^{\mathscr{A}}$ involved in the definition (6.7) of the vector functions $\varphi^{(i)}$, $i = 4, 5, \ldots, n = 3N + 3$, with the re-indexing (6.17).

6.4 Remark. The linear system used to calculate the approximations $\psi^{(n)}$ of $\psi = Tu$ and $Fa^{(n)}$ of Fa can be rewritten in a slightly more compact form. Thus, we designate $\bar{x}$ as $x^{(N+1)}$ and extend the notation $\varphi^{(jk)}$, $j = 1, 2, 3$, $k = N + 1$, to the 3 vector functions generated by (6.7) with $x^{(k)}$ replaced by $\bar{x} = x^{(N+1)}$; that is, (6.7) takes the augmented form

$$\varphi^{(jk)}(x) = \left((D(x, x^{(k)}))^{\mathscr{A}} \right)^{(j)}$$
$$= \left(\left((D(x^{(k)}, x))^{\mathscr{A}} \right)_{(j)} \right)^{\mathsf{T}}, \quad j = 1, 2, 3, \ k = 1, 2, \ldots, N + 1.$$

Accordingly, to the set $\{\varphi^{(i)}\}_{i=1}^{n}$ we add the vector functions

$$\varphi^{(n+i)} = \varphi^{(i, N+1)}, \quad i = 1, 2, 3.$$

Adopting this notational convention, we reformulate the computational procedure by seeking the approximation $\psi^{(n)}$ in the form

$$\psi^{(n)} = (Tu)^{(n)} = \sum_{h=1}^{n} c_h \varphi^{(h)}, \quad n = 3N + 3, \tag{6.26}$$

with the coefficients c_h determined from (6.22), (6.23), and (6.25) rewritten in the form

$$\sum_{h=1}^{n} c_h \langle f^{(i)}, \varphi^{(h)} \rangle = 0, \quad i = 1, 2, 3, \ n = 3N + 3, \tag{6.27}$$

$$\sum_{h=4}^{n} c_h \langle \varphi^{(i)}, \varphi^{(h)} \rangle - \left(F(x^{(k)}) \right)_{(j)} a^{(n)} = \left\langle \left(\left(P(x^{(k)}, \cdot) \right)_{(j)} \right)^{\mathsf{T}}, \mathscr{D} \right\rangle, \quad n = 3N + 3,$$

$$j = 1, 2, 3, \ k = 1, 2, \ldots, N+1, \ i = j + 3k = 4, 5, \ldots, 3N+6, \tag{6.28}$$

where we have set

$$a^{(n)} = (c_{n+1}, c_{n+2}, c_{n+3})^{\mathsf{T}} = (c_{3N+4}, c_{3N+5}, c_{3N+6})^{\mathsf{T}}. \tag{6.29}$$

The 3 equations (6.27) and $3N + 3$ equations (6.28) form our system for determining the $n + 3 = 3N + 6$ coefficients c_i. In expanded form, this system is

$$\begin{pmatrix} \langle f^{(1)}, \varphi^{(1)} \rangle & \langle f^{(1)}, \varphi^{(2)} \rangle & \cdots & \langle f^{(1)}, \varphi^{(n)} \rangle & 0 & 0 & 0 \\ \langle f^{(2)}, \varphi^{(1)} \rangle & \langle f^{(2)}, \varphi^{(2)} \rangle & \cdots & \langle f^{(2)}, \varphi^{(n)} \rangle & 0 & 0 & 0 \\ \langle f^{(3)}, \varphi^{(1)} \rangle & \langle f^{(3)}, \varphi^{(2)} \rangle & \cdots & \langle f^{(3)}, \varphi^{(n)} \rangle & 0 & 0 & 0 \\ \langle \varphi^{(4)}, \varphi^{(1)} \rangle & \langle \varphi^{(4)}, \varphi^{(2)} \rangle & \cdots & \langle \varphi^{(4)}, \varphi^{(n)} \rangle & -1 & 0 & 0 \\ \langle \varphi^{(5)}, \varphi^{(1)} \rangle & \langle \varphi^{(5)}, \varphi^{(2)} \rangle & \cdots & \langle \varphi^{(5)}, \varphi^{(n)} \rangle & 0 & -1 & 0 \\ \langle \varphi^{(6)}, \varphi^{(1)} \rangle & \langle \varphi^{(6)}, \varphi^{(2)} \rangle & \cdots & \langle \varphi^{(6)}, \varphi^{(n)} \rangle & x_1^{(1)} & x_2^{(1)} & -1 \\ \vdots & \vdots & & \vdots & \vdots & \vdots & \vdots \\ \langle \varphi^{(n+3)}, \varphi^{(1)} \rangle & \langle \varphi^{(n+3)}, \varphi^{(2)} \rangle & \cdots & \langle \varphi^{(n+3)}, \varphi^{(n)} \rangle & x_1^{(N+1)} & x_2^{(N+1)} & -1 \end{pmatrix} \begin{pmatrix} c_1 \\ c_2 \\ c_3 \\ \vdots \\ c_n \\ c_{n+1} \\ c_{n+2} \\ c_{n+3} \end{pmatrix}$$

$$= \begin{pmatrix} 0 \\ 0 \\ 0 \\ \left\langle \left(\left(P(x^{(1)}, \cdot) \right)_{(1)} \right)^{\mathsf{T}}, \mathscr{D} \right\rangle \\ \left\langle \left(\left(P(x^{(1)}, \cdot) \right)_{(2)} \right)^{\mathsf{T}}, \mathscr{D} \right\rangle \\ \left\langle \left(\left(P(x^{(1)}, \cdot) \right)_{(3)} \right)^{\mathsf{T}}, \mathscr{D} \right\rangle \\ \vdots \\ \left\langle \left(\left(P(x^{(N+1)}, \cdot) \right)_{(3)} \right)^{\mathsf{T}}, \mathscr{D} \right\rangle \end{pmatrix}. \tag{6.30}$$

6.5 Remark. We have deliberately written $f^{(i)}$ instead of $\varphi^{(i)}$ in the inner products in the first three rows in (6.30), to emphasize that those equations have a different source than the rest.

6.6 Remark. The additional point $\bar{x}$ in S^+ may be chosen either on or off ∂S_*. In the latter case, we must make sure that the augmented set $\{\varphi^{(i)}\}_{i=1}^{n+1}$ is linearly independent. According to Theorem 6.2, this property is satisfied if $\bar{x}$ is on the auxiliary curve. Under the above condition, system (6.30) is nonsingular, therefore, it has a unique solution, which generates the approximations (6.26) of $\psi^{(n)}$ and

$$(Fa)^{(n)} = Fa^{(n)}.$$

In our examples, we place $\bar{x}$ off ∂S_* to illustrate the viability of this option.

Using (6.3), we now define the vector functions

$$(u^{\mathscr{A}})^{(n)}(x) = -\langle (D(x,\cdot))^{\mathscr{A}}, \psi^{(n)} \rangle + \langle P(x,\cdot), \mathscr{D} \rangle, \quad x \in S^-, \tag{6.31}$$

and

$$u^{(n)}(x) = (u^{\mathscr{A}})^{(n)}(x) + F(x)a^{(n)}, \quad x \in S^-. \tag{6.32}$$

6.7 Remark. In the interior Dirichlet problem, the rigid displacement in the solution is computed separately. In the interior Neumann problem, it remains completely arbitrary and plays no role in the calculation. In both these cases, the linear algebraic system generated by the boundary data has a nonzero null space that needs special attention. Here, by contrast, the rigid displacement cannot be isolated from the full numerical computation process. The class $\mathscr{A}$ exterior Dirichlet problem (6.18), (6.19) together with the additional condition (6.24) give rise to a unique solution for $u = u^{\mathscr{A}} + Fa$ in S^-. This helps simplify the formulation of the algorithm and the computation.

With the c_i computed from (6.30), $a^{(n)}$ given by (6.29), and $(u^{\mathscr{A}})^{(n)}$ supplied by (6.31), approximation (6.32) takes the form

$$u^{(n)}(x) = -\langle (D(x,\cdot))^{\mathscr{A}}, \psi^{(n)} \rangle + \langle P(x,\cdot), \mathscr{D} \rangle + F(x)a^{(n)}, \quad x \in S^-. \tag{6.33}$$

6.8 Remark. In [15], Theorem 6.7, p. 140, it is shown that the rigid displacement Fa can be computed independently from $\mathscr{D}$ and the null space properties of the boundary operator $\frac{1}{2}I + W_0^*$ defined in Theorem 1.33(iii) (see [14, Theorem 3.22, p. 140]). In view of this, we may redefine $u^{(n)}$ alternatively as

$$u^{(n)}(x) = (u^{\mathscr{A}})^{(n)}(x) + (Fa)(x), \quad x \in S^-. \tag{6.34}$$

This form is preferred in analytic considerations, whereas (6.33) is the version of choice in computational work.

6.9 Theorem. *The vector function $u^{(n)}$ defined by (6.34) is an approximation of the solution u of problem (6.1) in the sense that $u^{(n)} \to u$ uniformly on any closed and bounded subdomain S' of S^-.*

Proof. By (6.2), (6.34), (6.3), and (6.31),

$$u(x) - u^{(n)}(x) = u^{\mathscr{A}}(x) - (u^{\mathscr{A}})^{(n)}(x)$$

$$= -\left\langle (D(x,\cdot))^{\mathscr{A}}, \psi \right\rangle + \left\langle (D(x,\cdot))^{\mathscr{A}}, \psi^{(n)} \right\rangle$$

$$= -\left\langle (D(x,\cdot))^{\mathscr{A}}, \psi - \psi^{(n)} \right\rangle, \quad x \in S^-.$$

Since

$$\lim_{n\to\infty} \| \psi - \psi^{(n)} \| = 0,$$

the argument now proceeds as in the proof of Theorem 3.6 in Sect. 3.2, with D replaced by $D^{\mathscr{A}}$. $\qquad\qquad\square$

6.1.2 Orthonormalization Method

After the re-indexing (6.16)–(6.17), we orthonormalize the set $\{\varphi^{(i)}\}_{i=1}^n$ as

$$\{\omega^{(i)}\}_{i=1}^n, \quad n = 3N+3.$$

Ordinarily, we would then try to compute the approximation $\psi^{(n)}$ of ψ by means of formula (2.2); that is,

$$\psi^{(n)} = \sum_{i=1}^n \langle \psi, \omega^{(i)} \rangle \omega^{(i)}$$

$$= \sum_{i=1}^n \left\langle \psi, \sum_{j=1}^i \bar{r}_{ij} \varphi^{(j)} \right\rangle \left(\sum_{h=1}^i \bar{r}_{ih} \varphi^{(h)} \right)$$

$$= \sum_{i=1}^n \sum_{j=1}^i \bar{r}_{ij} \langle \psi, \varphi^{(j)} \rangle \left(\sum_{h=1}^i \bar{r}_{ih} \varphi^{(h)} \right) \quad \text{on } \partial S,$$

where, by (6.24), $\langle \psi, \varphi^{(j)} \rangle = 0$ for $j = 1,2,3$, and the $\langle \psi, \varphi^{(j)} \rangle$ for $j = 4,5,\ldots,n$ are expected to be computed from (6.19) with x replaced by $x^{(k)}$ and $Tu^{\mathscr{A}} = \psi$; that is,

$$\left\langle (D(x^{(k)},\cdot))^{\mathscr{A}}, \psi \right\rangle = \langle P(x^{(k)},\cdot), \mathscr{D} \rangle, + (Fa)(x^{(k)})$$

or

$$\int_{\partial S} (D(x^{(k)},x))^{\mathscr{A}} \psi(x)\, ds(x) = \int_{\partial S} P(x^{(k)},x) \mathscr{D}(x)\, ds(x) + (Fa)(x^{(k)}),$$

which, by (6.7) and (6.17), yields

$$\langle \psi, \varphi^{(i)} \rangle = \int_{\partial S} \left((D(x^{(k)}))^{\mathscr{A}} \right)_{(j)} \psi(x) \, ds(x)$$

$$= \int_{\partial S} \left(P(x^{(k)}, x) \right) \mathscr{D}(x) \, ds(x) + (Fa)_j(x^{(k)}).$$

However, here we encounter a problem. This equality requires knowledge of Fa, which our method does not provide a priori and needs to be obtained as mentioned in Remark 6.8. Because of this, to stay exclusively within the confines of our procedure, we prefer to solve the exterior Dirichlet problem by means of the MRR method.

6.2 Numerical Example 1

Let S^+ be the disk of radius 1 centered at the origin, and suppose that, after rescaling and non-dimensionalization, the physical and geometric parameters of the plate are

$$h = 0.5, \quad \lambda = \mu = 1.$$

We choose the boundary condition, written in polar coordinates, in (6.1) to be

$$\mathscr{D}(x_p) = \begin{pmatrix} 1 + 2\cos(\theta) - 4\cos(3\theta) + 2\sin(4\theta) \\ -1 - 2\sin(\theta) - 2\cos(2\theta) - 4\sin(3\theta) - 2\cos(4\theta) \\ 2 - \cos\theta + 2\sin\theta - 9\cos(2\theta) + 7\sin(3\theta) \end{pmatrix}.$$

It is easily verified that the exact solution in S^- for this function is

$$u(x) = \begin{pmatrix} 1 + (x_1^2 + x_2^2)^{-1}(2x_1) \\ \quad + (x_1^2 + x_2^2)^{-2}(-2x_1^3 + 6x_1x_2^2) \\ \quad + (x_1^2 + x_2^2)^{-3}(-2x_1^3 + 6x_1x_2^2 + 8x_1^3x_2 - 8x_1x_2^3) \\ -1 + (x_1^2 + x_2^2)^{-1}(-2x_2) \\ \quad + (x_1^2 + x_2^2)^{-2}(-2x_1^2 + 2x_2^2 - 6x_1^2x_2 + 2x_2^3) \\ \quad + (x_1^2 + x_2^2)^{-3}(-6x_1^2x_2 + 2x_2^3 - 2x_1^4 + 12x_1^2x_2^2 - 2x_2^4) \\ 2 - x_1 + x_2 + (x_1^2 + x_2^2)^{-1}(x_2 - 2x_1^2 + 2x_2^2) \\ \quad + (x_1^2 + x_2^2)^{-2}(-7x_1^2 + 7x_2^2 + 3x_1^2x_2 - x_2^3) \\ \quad + (x_1^2 + x_2^2)^{-3}(18x_1^2x_2 - 6x_2^3) \end{pmatrix}, \quad (6.35)$$

or, in polar coordinates,

$$u(x_p) = \begin{pmatrix} 1 + r^{-1}[2\cos\theta - 2\cos(3\theta)] \\ \quad + r^{-2}[2\sin(4\theta)] + r^{-3}[-2\cos(3\theta)] \\ -1 + r^{-1}[-2\sin\theta - 2\sin(3\theta)] \\ \quad + r^{-2}[-2\cos(2\theta) - 2\cos(4\theta)] + r^{-3}[-2\sin(3\theta)] \\ r(-\cos\theta + \sin\theta) + 2 + r^{-1}[\sin\theta + \sin(3\theta)] \\ \quad + r^{-2}[-7\cos(2\theta)] + r^{-3}[6\sin(3\theta)] \end{pmatrix}.$$

6.10 Remark. Using this polar form and formulas (1.15), we easily convince ourselves that solution (6.35) belongs to $\mathscr{A}^*$, with the class $\mathscr{A}$ component $u^{\mathscr{A}}$ defined by the coefficients

$$\begin{aligned} m_0 &= 0, \\ m_1 &= 1, & m_2 &= -1, \\ m_3 &= -1, & m_4 &= 2, \\ m_5 &= 0, & m_6 &= 0, \\ m_7 &= 0, & m_8 &= -1, \\ m_9 &= 0, & m_{10} &= -1, \\ m_{11} &= 0, & m_{12} &= -7. \end{aligned}$$

The procedure described in Remark 1.11 also shows that this solution contains the rigid displacement

$$f = f^{(1)} - f^{(2)} + 2f^{(3)},$$

alternatively written as

$$f = Fa, \quad a = (1, -1, 2)^{\mathsf{T}}.$$

We select the auxiliary curve ∂S_* to be the circle of radius 1/2 centered at the origin.

6.11 Remark. This choice is based on the fact that

(i) if ∂S_* is too far away from ∂S, then the sequence $\mathscr{G}$ becomes 'less linearly independent';

(ii) if ∂S_* is too close to ∂S, then $\mathscr{G}$ becomes increasingly sensitive to the singularities of D and P on the boundary.

6.12 Remark. Clearly, the accuracy of the approximation depends on the selection of the set of points $\{x^{(k)}\}_{k=1}^n$. We make the reasonable choice of spacing these points uniformly around ∂S_*; that is, for $n = 1, 2, \ldots,$

$$\{x^{(k)} : k = 1, 2, \ldots, n\}_{\text{Cartesian}} = \left\{ \left(\frac{1}{2}, \frac{2\pi k}{n} \right) : k = 1, 2, \ldots, n \right\}_{\text{Polar}}.$$

6.13 Remark. Obviously, $\{x^{(k)}\}_{k=1}^\infty$ is the set of all points on ∂S_* whose polar angle is of the form $2\pi\alpha$, where α is any rational number such that $0 < \alpha \leq 1$.

6.14 Remark. We have performed floating-point computation with machine precision of approximately 16 digits. The most sensitive part of the process is the evaluation of integrals in the inner products, for which we set a target of 11 significant digits.

As already mentioned, we approximate the solution of problem (6.1) in this example by means of the MRR method.

6.2.1 Graphical Illustrations

We pointed out earlier (see Remark 6.10) that the exact solution u includes the rigid displacement component $f = f^{(1)} - f^{(2)} + 2f^{(3)}$, which is approximated as $f^{(n)}$ during the computation. As indicated at the beginning of this chapter, $f|_{\partial S}$ needs to be subtracted from $\mathscr{D}$ in the process of constructing the approximation of $u^{\mathscr{A}}$.

The graphs of the components of $(u^{\mathscr{A}})^{(63)}$ computed from $\psi^{(63)}$ (that is, with 20 points $x^{(k)}$ on ∂S_*) for

$$r \geq 1.01, \quad 0 \leq \theta < 2\pi,$$

together with the graph of $\mathscr{D} - f$ on ∂S, are shown in Fig. 6.1.

The reason for the restriction $r \geq 1.01$ is that the influence of the singularities of $D^{\mathscr{A}}(x,y)$ and $P(x,y)$ increases considerably for $x \in S^-$ very close to $y \in \partial S$.

Figure 6.2 shows the graphs of the components of $(u^{\mathscr{A}})^{(63)}$ computed from $\psi^{(63)}$ for

$$1.01 \leq r \leq 100, \quad 0 \leq \theta < 2\pi,$$

which illustrate the class $\mathscr{A}$ behavior of the solution away from the boundary. The vertical range of these plots has been truncated to allow better visualization.

The graphs of the components of the actual error $(u^{\mathscr{A}})^{(63)} - u^{\mathscr{A}}$, together with the 20 points $x^{(k)}$ on ∂S_* and the additional point $\bar{x} = \left(\frac{1}{4}, \frac{1}{4}\right)$ chosen for use in (6.22), are shown in Fig. 6.3. The approximation is 4–5 digits of accuracy near ∂S, and improves significantly away from the boundary.

6.2.2 Computational Procedure

The step-by-step outline of the computational procedure used for solving the exterior Dirichlet problem starts with the boundary data function $\mathscr{D}$, whose components (in polar coordinates) are graphed in Fig. 6.4.

$\mathscr{D}$ is digitized on ∂S with a sufficient number of points to guarantee that the accuracy of all quadrature calculations significantly exceeds the expected magnitude of the error produced by our approximation method. This allows the method error to be isolated from any quadrature errors. For this example, 100 discrete points on ∂S prove adequate to reach our objective. Next, the functions $\varphi^{(i)}$ are constructed as described in (6.16) and (6.17), and are digitized as well. If instead of MRR we were to select an orthonormalization method—say, MGS—with the rigid displacement computed a priori as explained in Remark 6.8, then the $\omega^{(i)}$ would be digitized in the same way.

The graphs of typical three-component vector functions $\varphi^{(i)}$ and $\omega^{(i)}$ are displayed in Figs. 6.5 and 6.6, respectively. The increasing number of oscillations in the latter eventually produces computational difficulties in the evaluation of the inner products occurring in the construction of additional orthonormal vector functions.

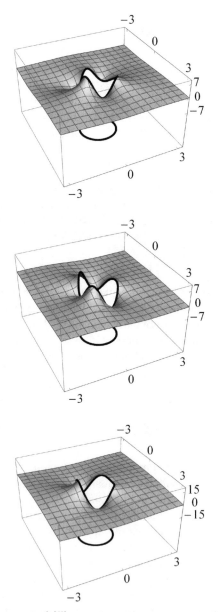

Fig. 6.1 Components of $(u^{\mathscr{A}})^{(63)}$ and $\mathscr{D} - f$ (heavy lines around the edge of the hole)

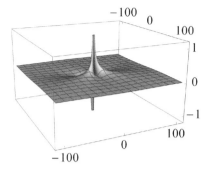

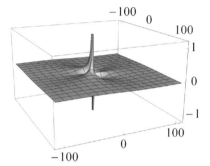

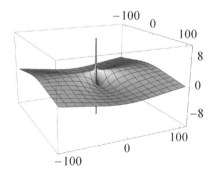

Fig. 6.2 Components of $(u^{\mathscr{A}})^{(63)}$ away from the origin

6.2.3 Computational Results

For the exterior Dirichlet problem, the approximation $\psi^{(n)}$ is computed by means of the system of equations generated by (6.4) and the $x^{(k)}$ and $\bar{x}$. In our example, $\psi^{(63)}$ (see Fig. 6.7) was constructed with 20 points $x^{(k)}$ on ∂S_* and the additional point $\bar{x} = \left(\frac{1}{4}, \frac{1}{4}\right)$. As Fig. 6.8 shows, this approximation is quite good when compared with the exact result for Tu derived from the known solution u.

Equation (6.32) and approximation $\psi^{(63)}$ generate the results for $u^{(63)}$ displayed in Figs. 6.1, 6.2, and 6.3.

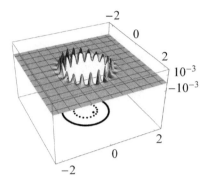

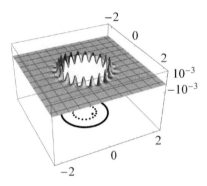

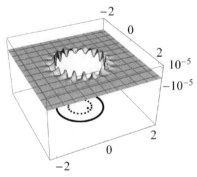

Fig. 6.3 Components of the error $(u^{\mathscr{A}})^{(63)} - u^{\mathscr{A}}$

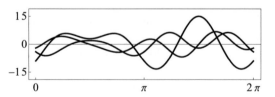

Fig. 6.4 Components of $\mathscr{D}$ in polar coordinates

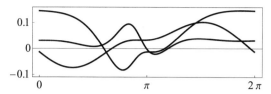

Fig. 6.5 Typical vector functions $\varphi^{(i)}$ in polar coordinates

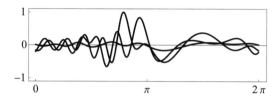

Fig. 6.6 Typical vector functions $\omega^{(i)}$ in polar coordinates

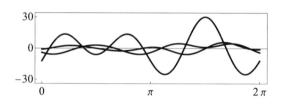

Fig. 6.7 Components of $\psi^{(63)}$ in polar coordinates

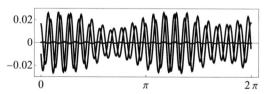

Fig. 6.8 Components of the error $\psi^{(63)} - Tu$ on ∂S in polar coordinates

6.2.4 Error Analysis

The approximation $\psi^{(n)}$, $n = 3N + 3$, is computed by means of N equally spaced points $x^{(k)}$ on ∂S_*, and the additional point $\bar{x}$. The behavior of the relative error

$$\frac{\|\psi^{(3N+3)} - Tu\|}{\|Tu\|}$$

as a function of N determines the efficiency and accuracy of the computational procedure. The logarithmic plot in Fig. 6.9 illustrates the size of the relative error as a function of N.

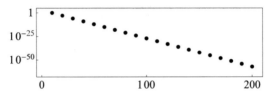

Fig. 6.9 The relative error $\|\psi^{(3N+3)} - Tu\|/\|Tu\|$ for $N = 5, 10, \ldots, 200$

This plot strongly suggests that the relative error improves exponentially as N increases. Fitting a linear curve to the logarithmic data produces the model

$$3.0867 - 0.301605\,N.$$

A comparison of the logarithmic data and the linear model is shown in Fig. 6.10.

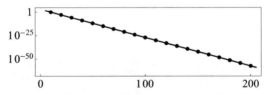

Fig. 6.10 The relative error $\|\psi^{(3N+3)} - Tu\|/\|Tu\|$ and the linear model for $N = 5, 10, \ldots, 200$

The two are seen to be in good agreement; consequently, the relative error may be modeled by

$$\frac{\|\psi^{(3N+3)} - Tu\|}{\|Tu\|} = 1221.09 \times 10^{-0.301605\,N} = 1221.09 \times 0.499339^N.$$

The exponentially decreasing error is optimal in the sense that it is superior to that of any fixed-order method.

Apart from forming the basis for the approximation $\psi^{(n)}$, representation formula (6.4) (also written as (6.19)) generates the approximation

$$f^{(n)} = a_1^{(n)} f^{(1)} + a_2^{(n)} f^{(2)} + a_3^{(n)} f^{(3)},$$

or

$$f^{(n)} = Fa^{(n)},$$

of the rigid displacement

$$f = Fa = f^{(1)} - f^{(2)} + 2f^{(3)}.$$

Here, we have used the notation introduced earlier, namely

$$a^{(n)} = (a_1^{(n)}, a_2^{(n)}, a_3^{(n)})^\mathsf{T} = (c_{n+1}, c_{n+2}, c_{n+3})^\mathsf{T}$$
$$= (c_{3N+4}, c_{3N+5}, c_{3N+6})^\mathsf{T}.$$

In our case, this approximation is

$$f^{(63)} = (1.00\ldots)f^{(1)} - (1.00\ldots)f^{(2)} + (2.00\ldots)f^{(3)},$$

with an error

$$f^{(63)} - f = -1.77 \times 10^{-8} f^{(1)} + 2.84 \times 10^{-8} f^{(2)} + 2.23 \times 10^{-9} f^{(3)}.$$

Figure 6.11 illustrates the relative error

$$\frac{\|a^{(n)} - a\|}{\|a\|}$$

in the computation of the constant vector $a = (1, -1, 2)^\mathsf{T}$.

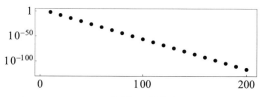

Fig. 6.11 The relative error $\|a^{(n)} - a\| / \|a\|$ for $N = 5, 10, \ldots, 200$

The plot shows that this relative error is much smaller than the corresponding relative error in the construction of $\psi^{(n)}$ as an approximation to Tu (see Fig. 6.10), which is to be expected, given that (6.4) is very error-sensitive to the procedure regarding the computation of $a^{(n)}$.

6.3 Numerical Example 2

With the same plate parameters, auxiliary circle ∂S_*, and points $x^{(k)}$ as in Sect. 6.2, we choose the boundary condition, written in polar coordinates, in (6.1) to be

$$
\mathcal{D}(x_p) =
\begin{pmatrix}
2 - \cos\theta + \sin\theta - 4\cos(2\theta) - 2\sin(2\theta) \\
\quad + \sin(3\theta) + 2\cos(4\theta) + 4\sin(4\theta) \\
1 + \cos\theta - \sin\theta - 2\cos(2\theta) - 2\sin(2\theta) \\
\quad - \cos(3\theta) - 4\cos(4\theta) + 2\sin(4\theta) \\
-2 - 5\cos\theta - \sin\theta - 3\sin(2\theta) + 7\cos(3\theta) \\
\quad + 14\sin(3\theta)
\end{pmatrix}.
\tag{6.36}
$$

The exact solution u in S^- of the problem is not known in this case.

The details mentioned in Remarks 6.11–6.14 apply here as well.

We approximate the solution of problem (6.1) in this example by means of the MRR method.

6.3.1 Graphical Illustrations

The computational procedure described in Sect. 6.1 shows that the approximation of the rigid displacement part of the solution of problem (6.1) with the boundary data (6.36) is

$$
f^{(n)} = (1.999\ldots)f^{(1)} + (1.000\ldots)f^{(2)} - (2.999\ldots)f^{(3)}.
\tag{6.37}
$$

According to what we said earlier, $f|_{\partial S}$ needs to be subtracted from $\mathcal{D}$ when the approximation of $u^{\mathscr{A}}$ is computed. The graphs of the components of $(u^{\mathscr{A}})^{(78)}$ generated by $\psi^{(78)}$ (that is, with 25 points $x^{(k)}$ on ∂S_*) for

$$
r \geq 1.01, \quad 0 \leq \theta < 2\pi,
$$

together with those of the components of $\mathcal{D} - f^{(78)}$, are shown in Fig. 6.12. The plots also show the points $x^{(k)}$ and the additional point $\bar{x} = \left(\frac{1}{4}, \frac{1}{4}\right)$ chosen as part of the computation of $\psi^{(78)}$.

The statement in Remark 6.14 is also valid for this example.

The reason for the restriction $r \geq 1.01$ is that the influence of the singularities of $D^{\mathscr{A}}(x,y)$ and $P(x,y)$ increases considerably for $x \in S^-$ very close to $y \in \partial S$.

Figure 6.13 shows the graphs of the components of $(u^{\mathscr{A}})^{(78)}$ computed from $\psi^{(78)}$ for

$$
1.01 \leq r \leq 100, \quad 0 \leq \theta < 2\pi,
$$

which illustrate the class $\mathscr{A}$ behavior of the solution away from the boundary. The vertical range of these plots has been truncated to allow better visualization.

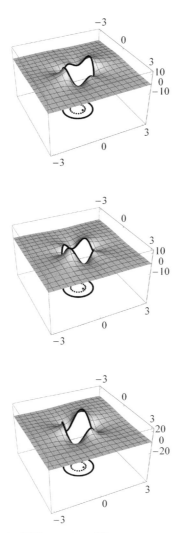

Fig. 6.12 Components of $(u^{\mathscr{A}})^{(78)}$ and $\mathscr{D} - f^{(78)}$ (heavy lines around the edge of the hole)

6.3.2 Computational Procedure

The step-by-step outline of the computational procedure used for solving the exterior Dirichlet problem starts with the boundary data function $\mathscr{D}$, whose components (in polar coordinates), including those of the rigid displacement, are graphed in Fig. 6.14.

$\mathscr{D}$ is now digitized on ∂S and the functions $\varphi^{(i)}$ are constructed and digitized as described in Sect. 6.2.2.

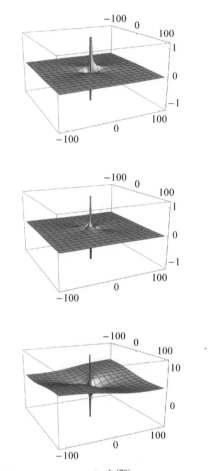

Fig. 6.13 Components of $(u^{\mathscr{A}})^{(78)}$ away from the origin

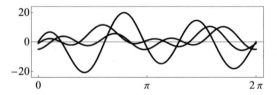

Fig. 6.14 Components of $\mathscr{D}$ in polar coordinates

6.3.3 Computational Results

For the exterior Dirichlet problem, the approximation $\psi^{(n)}$ of Tu (on ∂S) is computed by means of the system of equations generated by (6.4), the $x^{(k)}$, and $\bar{x}$. In our example, $\psi^{(78)}$ was constructed with 25 points $x^{(k)}$ on ∂S_* and the additional point $\bar{x} = (\frac{1}{4}, \frac{1}{4})$ (see Fig. 6.15). This approximation was then combined with the boundary condition function $\mathscr{D}$ and $f^{(78)}$ to generate the graphs in Fig. 6.12, which show good agreement between $(u^{\mathscr{A}})^{(78)}$ and $\mathscr{D} - f^{(78)}$. This is an indirect validation of our method.

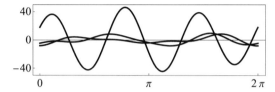

Fig. 6.15 Components of $\psi^{(78)}$ in polar coordinates

In addition to producing $\psi^{(78)}$, the representation formula (6.22) also yielded the approximation

$$f^{(78)} = (2.00\ldots)f^{(1)} + (1.00\ldots)f^{(2)} - (3.00\ldots)f^{(3)}.$$

When the rigid displacement f is computed with 200 points $x^{(k)}$, the error for $f^{(78)}$ can be approximated by

$$f^{(78)} - f^{(603)} = -6.20 \times 10^{-11} f^{(1)} + 7.04 \times 10^{-11} f^{(2)} + 1.80 \times 10^{-12} f^{(3)}.$$

6.3.4 Error Analysis

The approximation $\psi^{(n)}$, $n = 3N+3$, is computed with N equally spaced points $x^{(k)}$ on ∂S_*, and the additional point $\bar{x}$. The relative error

$$\frac{\|\psi^{(3N+3)} - Tu\|}{\|Tu\|}$$

as a function of N determines the efficiency and accuracy of the computational procedure. However, since the exact expression of Tu is not known in this example, an accurate error analysis cannot be performed.

As a substitute, we propose instead an indirect route to obtaining a measure of partial validation of our calculations. We use the approximation $\psi^{(78)}$ of Tu on ∂S as the boundary condition for an exterior Neumann problem and compute the approximate boundary value $\bar{u}^{(78)}|_{\partial S}$ of the solution to the latter by means of the scheme designed in Chap. 7. If our method is sound, then $\bar{u}^{(78)}|_{\partial S}$ augmented with $f^{(78)}$ given by (6.37) should be close to $\mathscr{D}$ (see Sect. 6.3.1). Figure 6.16 shows the components of $\bar{u}^{(78)}|_{\partial S} + f^{(78)}$.

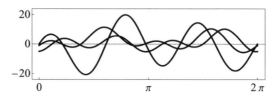

Fig. 6.16 Components of $\bar{u}^{(78)}|_{\partial S} + f^{(78)}$ in polar coordinates

The components of the error $\bar{u}^{(78)}|_{\partial S} + f^{(78)} - \mathscr{D}$ are shown in Fig. 6.17.

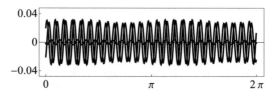

Fig. 6.17 Components of $\bar{u}^{(78)}|_{\partial S} + f^{(78)} - \mathscr{D}$ in polar coordinates

This plot indicates that the total combined error in this verification exercise is consistent with the error expected when we use MRR to compute $u^{(78)}|_{\partial S}$. Repeating our numerical experiment with 200 points $x^{(k)}$ instead of 25 on ∂S_*, we obtained the much smaller relative error

$$\frac{\|\bar{u}^{(603)}|_{\partial S} + f^{(603)} - \mathscr{D}\|}{\|\mathscr{D}\|} = 6.27934 \times 10^{-57}.$$

This confirms the robust nature of the generalized Fourier series method of approximation presented here.

Chapter 7
Exterior Neumann Problem

7.1 Computational Algorithm

The exterior Neumann problem consists in finding u such that

$$Au(x) = 0, \quad x \in S^-,$$
$$Tu(x) = \mathcal{N}(x), \quad x \in \partial S, \tag{7.1}$$
$$u \in \mathcal{A},$$

where the vector function $\mathcal{N} \in C^{(0,\alpha)}(\partial S)$, $\alpha \in (0,1)$, is prescribed.

Since the solution u of (7.1) is of class $\mathcal{A}$, and in keeping with the notation used in the exterior Dirichlet and Robin problems, throughout this chapter we will refer to it as $u^{\mathcal{A}}$.

The representation formulas hold for a solution of the system in S^- that belongs to the asymptotic class $\mathcal{A}$, so, by (1.22),

$$u^{\mathcal{A}}(x) = -\int_{\partial S} D(x,y)\mathcal{N}(y)\,ds(y) + \int_{\partial S} P(x,y)u^{\mathcal{A}}(y)\,ds(y), \quad x \in S^-, \tag{7.2}$$

$$0 = -\int_{\partial S} D(x,y)\mathcal{N}(y)\,ds(y) + \int_{\partial S} P(x,y)u^{\mathcal{A}}(y)\,ds(y), \quad x \in S^+. \tag{7.3}$$

Let ∂S_* be a simple, closed, C^2–curve lying strictly inside S^+, let $\{x^{(k)}\}_{k=1}^\infty$ be a set of points densely distributed on ∂S_*, and consider the sequence of vector functions on ∂S defined by

$$\mathcal{G} = \{\varphi^{(jk)}, \ j = 1,2,3, \ k = 1,2,\ldots\},$$

where, by (1.2),

$$\varphi^{(jk)}(x) = \left(T\left(D(x,x^{(k)})\right)^{\mathcal{A}}\right)^{(j)}$$
$$= T\left(\left(D(x,x^{(k)})\right)^{\mathcal{A}}\right)^{(j)}, \quad j = 1,2,3, \ k = 1,2,\ldots. \tag{7.4}$$

© Springer Nature Switzerland AG 2020

C. Constanda, D. Doty, *The Generalized Fourier Series Method*, Developments in Mathematics 65, https://doi.org/10.1007/978-3-030-55849-9_7

7.1 Remark. There is an alternative expression for $\varphi^{(jk)}$, written in terms of the matrix P. By (1.10) and the fact that $(P(x,y))^{\mathscr{A}}$ is constructed from $(D(x,y))^{\mathscr{A}}$ (see Sect. 1.2.2),

$$\left(P(x,y)\right)_{(j)} = \left((P(x,y))^{\mathscr{A}}\right)_{(j)} = \left((T(\partial_y)(D(y,x))^{\mathscr{A}})^{\mathsf{T}}\right)_{(j)}$$
$$= \left((T(\partial_y)(D(y,x))^{\mathscr{A}})^{(j)}\right)^{\mathsf{T}}, \qquad (7.5)$$

so

$$\left((P(x,y))_{(j)}\right)^{\mathsf{T}} = \left(T(\partial_y)(D(y,x))^{\mathscr{A}}\right)^{(j)},$$

which, in view of (7.4) and (6.5), yields

$$\varphi^{(jk)}(x) = \left(T(D(x,x^{(k)}))^{\mathscr{A}}\right)^{(j)} = \left((P(x^{(k)},x))_{(j)}\right)^{\mathsf{T}}, \quad x \in \partial S. \qquad (7.6)$$

7.2 Theorem. *The set $\mathscr{G}$ is linearly independent on ∂S and complete in $L^2(\partial S)$.*

Proof. Assuming the opposite, suppose that there are a positive integer N and real numbers c_{jk}, $i, j = 1, 2, 3$, $k = 1, 2, \ldots, N$, not all zero, such that

$$\sum_{j=1}^{3}\sum_{k=1}^{N} c_{jk}\varphi^{(jk)}(x) = 0, \quad x \in \partial S. \qquad (7.7)$$

We set

$$g(x) = \sum_{j=1}^{3}\sum_{k=1}^{N} c_{jk}\left((D(x,x^{(k)}))^{\mathscr{A}}\right)^{(j)}, \quad x \in \bar{S}^{-},$$

and see that, by (7.4), (7.7), and Theorem 1.9 applied to $D^{\mathscr{A}}$,

$$Ag = 0 \quad \text{in } S^{-},$$
$$Tg = 0 \quad \text{on } \partial S,$$
$$g \in \mathscr{A};$$

that is, g is the (unique) regular solution of the homogeneous Neumann problem in S^{-}. According to Theorem 1.29(v),

$$g = 0 \quad \text{in } \bar{S}^{-}. \qquad (7.8)$$

But $\left|\left((D(x,x^{(k)}))^{\mathscr{A}}\right)_{3j}\right| \to \infty$ as $|x| \to \infty$ since it contains a $\ln|x|$ term for every $j = 1, 2, 3$ and $k = 1, 2, \ldots, N$, which, therefore, is also present in g_3. This contradicts (7.8), so all $c_{jk} = 0$, proving the linear independence of $\mathscr{G}$.

For the second part of the proof, suppose that there is $q \in L^2(\partial S)$ such that

$$\langle q, \varphi^{(jk)}\rangle = \int_{\partial S}(\varphi^{(jk)})^{\mathsf{T}} q\, ds = 0, \quad j = 1,2,3, \ k = 1,2,\ldots. \qquad (7.9)$$

By (7.6), equality (7.9) is equivalent to

$$\int_{\partial S} \left(T\left(D(x,x^{(k)})\right)^{\mathscr{A}}\right)^{\mathsf{T}} q(x)\, ds = \int_{\partial S} P(x^{(k)},x) q(x)\, ds = 0, \quad k = 1, 2, \ldots.$$

Since the points $x^{(k)}$ are in S^+, the double-layer potential with an L^2-density

$$(Wq)(x) = \int_{\partial S} P(x,y) q(y)\, ds(y), \quad x \in S^+ \cup S^-,$$

satisfies

$$(Wq)(x^{(k)}) = \int_{\partial S} P(x^{(k)}(x)) q(x)\, ds = 0,$$

so, since the set $\{x^{(k)}\}_{k=1}^{\infty}$ is dense on ∂S_*,

$$Wq = 0 \quad \text{on } \partial S_*.$$

Hence,

$$A(Wq) = 0 \quad \text{in } S_*^+,$$
$$Wq = 0, \quad \text{on } \partial S_*,$$

which means that Wq is the (unique) solution of the homogeneous Dirichlet problem in S_*^+; therefore, $Wq = 0$ in $\bar{S}_*^+$ and, by analyticity,

$$Wq = 0 \quad \text{in } S^+.$$

By Theorem 4.20, p. 99 in [15], we then have

$$-\tfrac{1}{2} q(x) + \int_{\partial S} P(x,y) q(y)\, ds(y) = 0 \quad \text{for a.a. } x \in \partial S. \tag{7.10}$$

By Theorem 6.12, p.145 in [15], this implies that $q \in C^{0,\alpha}(\partial S)$, so (7.10) holds at every point on ∂S. Consequently, (7.10) is equivalent to

$$\left(W_0 - \tfrac{1}{2} I\right) = 0 \quad \text{on } \partial S$$

and so, by Theorem 3.22, p. 140 in [14], $q = 0$, which proves that $\mathscr{G}$ is a complete set in $L^2(\partial S)$. $\qquad\square$

We order the elements of $\mathscr{G}$ as the sequence

$$\varphi^{(11)}, \varphi^{(21)}, \varphi^{(31)}, \varphi^{(12)}, \varphi^{(22)}, \varphi^{(32)}, \varphi^{(13)}, \varphi^{(23)}, \varphi^{(33)}, \ldots$$

and re-index them, noticing that for each $i = 1, 2, \ldots$, there is a unique pair $\{j, k\}$, $j = 1, 2, 3$, $k = 1, 2, \ldots$, such that $i = j + 3(k-1)$. Then

$$\mathscr{G} = \{\varphi^{(1)}, \varphi^{(2)}, \varphi^{(3)}, \varphi^{(4)}, \varphi^{(5)}, \varphi^{(6)}, \varphi^{(7)}, \varphi^{(8)}, \varphi^{(9)}, \ldots\}, \tag{7.11}$$

where

$$\varphi^{(i)} = \varphi^{(jk)}, \quad j = 1,2,3, \ k = 1,2,\dots, \ i = j+3(k-1) = 1,2,\dots. \tag{7.12}$$

7.1.1 Row Reduction Method

Our intention is to construct a nonsingular linear system of finitely many algebraic equations to approximate the solution u of (7.1).

Rewriting the representation formulas (7.2) and (7.3) as

$$u^{\mathscr{A}}(x) = -\langle D(x,\cdot), \mathscr{N}\rangle + \langle P(x,\cdot), u^{\mathscr{A}}\rangle, \quad x \in S^-, \tag{7.13}$$

$$\langle P(x,\cdot), u^{\mathscr{A}}\rangle = \langle D(x,\cdot), \mathscr{N}\rangle, \quad x \in S^+, \tag{7.14}$$

we aim to use (7.14) to approximate $u^{\mathscr{A}}|_{\partial S}$, and then to approximate $u^{\mathscr{A}}$ in S^- by means of (7.13).

We set

$$\psi = u^{\mathscr{A}}|_{\partial S}.$$

Since the $x^{(k)}$ are points in S^+ and $\langle \mathscr{N}, f^{(i)}\rangle = 0$, $i = 1,2,3$, from (7.3) it follows that

$$\int_{\partial S} P(x^{(k)},x)\psi(x)\,ds(x) = \int_{\partial S} D(x^{(k)},x)\mathscr{N}(x)\,ds(x)$$

$$= \int_{\partial S} \left(D(x^{(k)},x)\right)^{\mathscr{A}}\mathscr{N}(x)\,ds(x),$$

which, by (7.6) and (7.12), leads to

$$\langle \varphi^{(i)}, \psi\rangle = \langle \varphi^{(jk)}, \psi\rangle = \int_{\partial S} \left(P(x^{(k)},x)\right)_{(j)}\psi(x)\,ds(x)$$

$$= \int_{\partial S} \left(\left(D(x^{(k)},x)\right)^{\mathscr{A}}\right)_{(j)}\mathscr{N}(x)\,ds(x),$$

$$j = 1,2,3, \ k = 1,2,\dots, \ i = j+3(k-1) = 1,2,\dots,$$

or, what is the same,

$$\langle \varphi^{(i)}, \psi\rangle = \langle \left(P(x^{(k)},\cdot)\right)_{(j)}, \psi\rangle = \langle \left(\left(D(x^{(k)},\cdot)\right)_{(j)}\right)^{\mathsf{T}}, \mathscr{N}\rangle. \tag{7.15}$$

In view of Theorem 7.2, there is a unique expansion

$$\psi = \sum_{h=1}^{\infty} c_h \varphi^{(h)}, \tag{7.16}$$

so we can seek an approximation of ψ of the form

$$\psi^{(n)} = (u^{\mathscr{A}}|_{\partial S})^{(n)} = \sum_{h=1}^{n} c_h \varphi^{(h)}. \tag{7.17}$$

Replacing ψ by $\psi^{(n)}$ in (7.15), for $i = j + 3(k-1) = 1, 2, \ldots, n$ we arrive at the approximate equality

$$\sum_{h=1}^{n} c_h \langle \varphi^{(i)}, \varphi^{(h)} \rangle = \left\langle \left(\left((D(x^{(k)}, x))^{\mathscr{A}} \right)_{(j)} \right)^{\mathsf{T}}, \mathscr{N} \right\rangle. \tag{7.18}$$

7.3 Remark. For symmetry and ease of computation, in the construction of the approximation $\psi^{(n)}$ we use the subsequence with $n = 3N$, where N is the number of points $x^{(k)}$ chosen on ∂S_*. This choice feeds into the computation, for each $k = 1, 2, \ldots, N$, all three rows of P involved in the definition (7.6) of the vector functions $\varphi^{(i)}$, $i = 1, 2, \ldots, n = 3N$, with the re-indexing (7.12).

In expanded form, (7.18) is

$$\begin{pmatrix} \langle \varphi^{(1)}, \varphi^{(1)} \rangle & \langle \varphi^{(1)}, \varphi^{(2)} \rangle & \cdots & \langle \varphi^{(1)}, \varphi^{(n)} \rangle \\ \langle \varphi^{(2)}, \varphi^{(1)} \rangle & \langle \varphi^{(2)}, \varphi^{(2)} \rangle & \cdots & \langle \varphi^{(2)}, \varphi^{(n)} \rangle \\ \vdots & \vdots & & \vdots \\ \langle \varphi^{(n)}, \varphi^{(1)} \rangle & \langle \varphi^{(n)}, \varphi^{(2)} \rangle & \cdots & \langle \varphi^{(n)}, \varphi^{(n)} \rangle \end{pmatrix} \begin{pmatrix} c_1 \\ c_2 \\ \vdots \\ c_n \end{pmatrix}$$

$$= \begin{pmatrix} \left\langle \left(\left((D(x^{(1)}, \cdot))^{\mathscr{A}} \right)_{(1)} \right)^{\mathsf{T}}, \mathscr{N} \right\rangle \\ \left\langle \left(\left((D(x^{(1)}, \cdot))^{\mathscr{A}} \right)_{(2)} \right)^{\mathsf{T}}, \mathscr{N} \right\rangle \\ \vdots \\ \left\langle \left(\left((D(x^{(N)}, \cdot))^{\mathscr{A}} \right)_{(3)} \right)^{\mathsf{T}}, \mathscr{N} \right\rangle \end{pmatrix}. \tag{7.19}$$

System (7.19) is nonsingular, therefore, it determines the $n = 3N$ coefficients c_h uniquely, which, in turn, generate the approximation (7.17) of $\psi^{(n)}$.

Using (7.2), for $x \in S^-$ we now define the vector function

$$u^{(n)}(x) = (u^{\mathscr{A}})^{(n)}(x)$$

$$= -\int_{\partial S} (D(x, y))^{\mathscr{A}} \mathscr{N}(y) \, ds(y) + \int_{\partial S} P(x, y) \psi^{(n)}(y) \, ds(y)$$

$$= -\langle (D(x, \cdot))^{\mathscr{A}}, \mathscr{N} \rangle + \langle P(x, \cdot), \psi^{(n)} \rangle. \tag{7.20}$$

7.4 Theorem. *The vector function $u^{(n)}$ defined by (7.20) is an approximation of the solution u of problem (7.1) in the sense that $u^{(n)} \to u$ uniformly on any closed and bounded subdomain S' of S^-.*

The proof of this assertion is identical to that of Theorem 6.9 in Sect. 6.1, with $D^{\mathscr{A}}$ replaced by P and no rigid displacement involvement.

7.5 Remark. In the interior Dirichlet problem, the rigid displacement in the solution is computed separately. In the interior Neumann problem, it remains completely arbitrary and plays no role in the calculation. In both these cases, the linear algebraic system generated by the boundary data has a nonzero null space that needs special attention. Here, by contrast, no rigid displacement is involved in the numerical computation process. The exterior Neumann problem (7.13), (7.14) gives rise to a unique solution for $\psi = u^{\mathscr{A}}|_{\partial S}$. This significantly simplifies the formulation of the algorithm and the computation.

7.6 Remark. The exterior Neumann problem has the necessary and sufficient condition of solvability

$$\langle \mathscr{N}, f^{(i)} \rangle = 0, \quad i = 1, 2, 3,$$

so the solution does not contain any rigid displacement. This is why the set $\mathscr{G}$ does not include the functions $f^{(i)}$, $i = 1, 2, 3$.

7.1.2 Orthonormalization Method

After the re-indexing (7.12), we orthonormalize the set $\{\varphi^{(i)}\}_{i=1}^{n}$ as

$$\{\omega^{(i)}\}_{i=1}^{n}, \quad n = 3N.$$

This yields the QR factorization

$$\Phi = \Omega R,$$

where Φ, Ω, and the nonsingular, upper-triangular $n \times n$ matrix R are described in Sect. 2.2.

We rewrite system (7.19) in the form

$$Mc = \beta,$$

where

$$M = (M_{ij}) = (\langle \varphi^{(i)}, \varphi^{(j)} \rangle), \quad 1 \le i, j \le n = 3N,$$
$$c = (c_1, c_2, \ldots, c_n)^{\mathsf{T}},$$

and β is the $3N$-column vector of the numbers on the right-hand side in (7.19).

According to the procedure outlined in Sect. 3.1.2, the approximation $\psi^{(n)}$ of $u^{\mathscr{A}}$ becomes

$$\psi^{(n)} = \Omega (R^{\mathsf{T}})^{-1} \beta. \tag{7.21}$$

Since R is upper triangular, the matrix

$$(R^{\mathsf{T}})^{-1} = (\bar{r}_{ih})$$

is lower triangular and can be computed by forward substitution. This reduces (7.21) to

$$\psi^{(n)} = \sum_{i=1}^{n} \left(\sum_{h=1}^{i} \bar{r}_{ih}\beta_h \right) \omega^{(i)}.$$

Once $\psi^{(n)}$ has been determined, we can use (7.20) to construct

$$u^{(n)} = (u^{\mathscr{A}})^{(n)}.$$

Remark 3.7 applies to this calculation as well.

7.2 Numerical Example 1

Let S^+ be the disk of radius 1 centered at the origin, and suppose that, after rescaling and non-dimensionalization, the physical and geometric parameters of the plate are

$$h = 0.5, \quad \lambda = \mu = 1.$$

We choose the boundary condition, written in polar coordinates, in (7.1) to be

$$\mathscr{N}(x_p) = \begin{pmatrix} 2\sin\theta - 6\cos(2\theta) + 3\cos(3\theta) + 3\sin(3\theta) + 3\cos(4\theta) \\ 2\cos\theta + 2\sin\theta + 6\sin(2\theta) - 3\cos(3\theta) + 3\sin(3\theta) + 3\sin(4\theta) \\ 6\cos(2\theta) - 12\sin(2\theta) + 54\cos(3\theta) \end{pmatrix}.$$

It is easily verified that this function satisfies the solvability condition (1.23), and that the exact solution in S^- generated by it is

$$u(x) = \begin{pmatrix} (x_1^2 + x_2^2)^{-1}[x_1 + 2x_2] \\ \quad + (x_1^2 + x_2^2)^{-2}[-3x_1^2 + x_1^3 - 6x_1^2 x_2 + 3x_2^2 - 3x_1 x_2^2 + 2x_2^3] \\ \quad + (x_1^2 + x_2^2)^{-3}[2x_1^3 + 6x_1^4 + 6x_1^2 x_2 - 6x_1 x_2^2 - 36x_1^2 x_2^2 \\ \qquad\qquad - 2x_2^3 + 6x_2^4] \\ (x_1^2 + x_2^2)^{-1}[2x_1 + 3x_2] \\ \quad + (x_1^2 + x_2^2)^{-2}[2x_1^3 + 6x_1 x_2 + 3x_1^2 x_2 - 6x_1 x_2^2 - x_2^3] \\ \quad + (x_1^2 + x_2^2)^{-3}[-2x_1^3 + 6x_1^2 x_2 + 24x_1^3 x_2 + 6x_1 x_2^2 \\ \qquad\qquad - 2x_2^3 - 24x_1 x_2^3] \\ (x_1^2 + x_2^2)^{-1}[x_1^2 - 4x_1 x_2 - x_2^2] \\ \quad + (x_1^2 + x_2^2)^{-2}[4x_1^2 + 3x_1^3 - 10x_1 x_2 - 4x_2^2 - 9x_1 x_2^2] \\ \quad + (x_1^2 + x_2^2)^{-3}[18x_1^3 - 54x_1 x_2^2] \end{pmatrix},$$

or, in polar coordinates,

$$u(x_p) = \begin{pmatrix} r^{-1}[\cos\theta + 2\sin\theta + \cos(3\theta) - 2\sin(3\theta)] \\ \quad + r^{-2}[-3\cos(2\theta) + 6\cos(4\theta)] \\ \quad\quad + r^{-3}[2\cos(3\theta) + 2\sin(3\theta)] \\ r^{-1}[2\cos\theta + 3\sin\theta + 2\cos(3\theta) + \sin(3\theta)] \\ \quad + r^{-2}[3\sin(2\theta) + 6\sin(4\theta)] \\ \quad\quad + r^{-3}[-2\cos(3\theta) + 2\sin(3\theta)] \\ r^{-1}[3\cos(3\theta)] \\ \quad + r^{-2}[4\cos(2\theta) - 5\sin(2\theta)] \\ \quad\quad + r^{-3}[18\cos(3\theta)] \end{pmatrix}.$$

7.7 Remark. Using this polar form and formulas (1.15), we easily convince our-selves that solution (7.22) belongs to class $\mathscr{A}$, with the m coefficients

$$m_0 = 2,$$
$$m_1 = \tfrac{1}{2}, \qquad m_2 = \tfrac{3}{2},$$
$$m_3 = 0, \qquad m_4 = 0,$$
$$m_5 = -3, \qquad m_6 = 3,$$
$$m_7 = 1, \qquad m_8 = 1,$$
$$m_9 = 1, \qquad m_{10} = 1,$$
$$m_{11} = -5, \qquad m_{12} = 4.$$

Also, as expected, the procedure described in Remark 1.11 shows that this solution has no rigid displacement component.

We select the auxiliary curve ∂S_* to be the circle of radius 1/2 centered at the origin.

The details mentioned in Remarks 6.11–6.14 apply here as well.

We approximate the solution of problem (7.1) in this example by means of the MGS method.

7.2.1 Graphical Illustrations

The graphs of the components of $(u^{\mathscr{A}})^{(75)}$ computed from $\psi^{(75)}$ (that is, with 25 points $x^{(k)}$ on ∂S_*) for

$$r \geq 1.01, \quad 0 \leq \theta < 2\pi,$$

together with those of $\psi^{(75)}$, which approximates $u^{\mathscr{A}}|_{\partial S}$, are shown in Fig. 7.1.

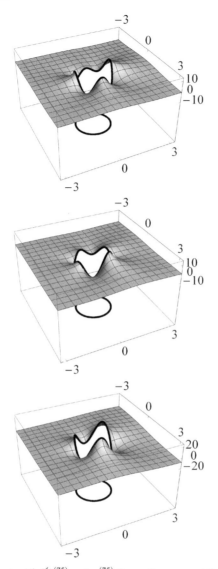

Fig. 7.1 Components of $(u^{\mathscr{A}})^{(75)}$ and $\psi^{(75)}$ (heavy lines around the edge of the hole)

The reason for $r \geq 1.01$ is that the influence of the singularities of $(D(x,y))^{\mathscr{A}}$ and $P(x,y)$ increases considerably for $x \in S^{-}$ very close to $y \in \partial S$.

Figure 7.2 shows the graphs of the components of $(u^{\mathscr{A}})^{(75)}$ computed from $\psi^{(75)}$ for

$$1.01 \leq r \leq 100, \quad 0 \leq \theta < 2\pi,$$

which illustrate the class $\mathscr{A}$ behavior of the solution away from the boundary. The vertical range of these plots has been truncated to allow better visualization.

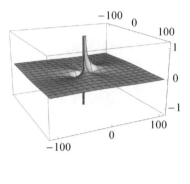

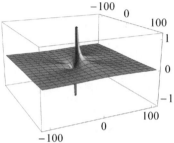

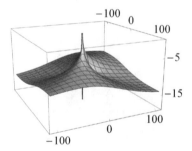

Fig. 7.2 Components of $(u^{\mathscr{A}})^{(75)}$ away from the origin

The graphs of the components of the actual error $(u^{\mathscr{A}})^{(75)} - u^{\mathscr{A}}$, together with ∂S and the 25 points $x^{(k)}$ on ∂S_*, are shown in Fig. 7.3. The approximation is 4–5 digits of accuracy near ∂S, and improves significantly away from the boundary.

7.2.2 Computational Procedure

The step-by-step outline of the computational procedure used for solving the exterior Neumann problem starts with the boundary data function $\mathscr{N}$, whose components (in polar coordinates) are graphed in Fig. 7.4.

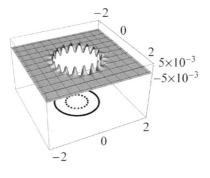

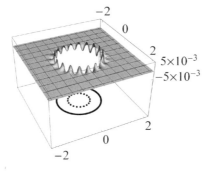

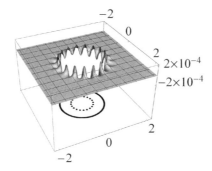

Fig. 7.3 Components of the error $(u^{\mathscr{A}})^{(75)} - u^{\mathscr{A}}$

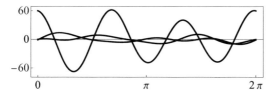

Fig. 7.4 Components of $\mathscr{N}$ in polar coordinates

$\mathcal{N}$ is digitized on ∂S with a sufficient number of points to guarantee that the accuracy of all quadrature calculations significantly exceeds the expected magnitude of the error produced by our approximation method. This allows the method error to be isolated from any quadrature errors. For this example, 100 discrete points on ∂S prove adequate to reach our objective. Next, the functions $\varphi^{(i)}$ are constructed as described in (7.4) and (7.12), and are digitized as well. An orthonormalization method is then selected—in our case, MGS—which generates the digitized $\omega^{(i)}$ as described earlier.

The graphs of typical three-component vector functions $\varphi^{(i)}$ and $\omega^{(i)}$ are displayed in Figs. 7.5 and 7.6, respectively. The increasing number of oscillations in the latter eventually produces computational difficulties in the evaluation of the inner products occurring in the construction of additional orthonormal vector functions.

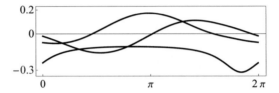

Fig. 7.5 Typical vector functions $\varphi^{(i)}$ in polar coordinates

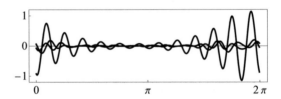

Fig. 7.6 Typical vector functions $\omega^{(i)}$ in polar coordinates

7.2.3 Computational Results

For the exterior Neumann problem, the approximation $\psi^{(n)}$ of $u^{\mathscr{A}}|_{\partial S}$ is computed by means of the system of equations generated by (7.4) and the $x^{(k)}$. In our example, $\psi^{(75)}$ (see Fig. 7.7) was constructed with 25 points $x^{(k)}$ on ∂S_*. As Fig. 7.8 shows, this approximation is quite good when compared with the exact result for $u^{\mathscr{A}}|_{\partial S}$ derived from the known solution $u^{\mathscr{A}}$.

Equation (7.2) and $\psi^{(75)}$ generate the results for $(u^{\mathscr{A}})^{(75)}$ displayed in Figs. 7.1, 7.2, 7.3.

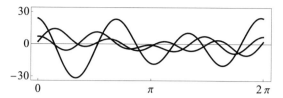

Fig. 7.7 Components of $\psi^{(75)}$ in polar coordinates

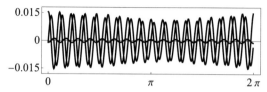

Fig. 7.8 Components of the error $\psi^{(75)} - u^{\mathscr{A}}|_{\partial S}$ in polar coordinates

7.2.4 Error Analysis

The approximation $\psi^{(n)}$, $n = 3N$, is computed using N equally spaced points $x^{(k)}$ on ∂S_*. The behavior of the relative error

$$\frac{\| \psi^{(3N)} - u^{\mathscr{A}}|_{\partial S} \|}{\| u^{\mathscr{A}}|_{\partial S} \|}$$

as a function of N determines the efficiency and accuracy of the computational procedure. The logarithmic plot in Fig. 7.9 illustrates the size of the relative error in terms of N.

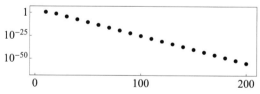

Fig. 7.9 The relative error $\| \psi^{(3N)} - u^{\mathscr{A}}|_{\partial S} \| / \| u^{\mathscr{A}}|_{\partial S} \|$ for $N = 5, 10, \ldots, 200$

This plot strongly suggests that the relative error improves exponentially as N increases. Fitting a linear curve to the logarithmic data produces the model

$$4.66474 - 0.297786 N.$$

A comparison of the logarithmic data and the linear model is shown in Fig. 7.10. The two are seen to be in good agreement, so the relative error may be modeled by

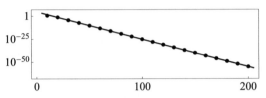

Fig. 7.10 The relative error $\|\psi^{(3N)} - u^{\mathscr{A}}|_{\partial S}\| / \|u^{\mathscr{A}}|_{\partial S}\|$ and the linear model for $N = 5, 10, \ldots, 200$

$$\frac{\|\psi^{(3N)} - u^{\mathscr{A}}|_{\partial S}\|}{\|u^{\mathscr{A}}|_{\partial S}\|} = 46210.4 \times 10^{-0.297886N}$$

$$= 46210.4 \times 0.503633^N.$$

The exponentially decreasing error is optimal in the sense that it is superior to that of any fixed-order method.

7.3 Numerical Example 2

With the same auxiliary circle ∂S_* and points $x^{(k)}$ as in Sect. 7.2, we choose the boundary condition, written in polar coordinates, in (7.1) to be

$$\mathscr{N}(x_p) = \begin{pmatrix} -\frac{5}{2}\cos\theta + 8\cos(2\theta) - 2\sin(2\theta) - 3\cos(3\theta) + 3\sin(3\theta) \\ \qquad\qquad - 3\cos(4\theta) + \sin(4\theta) \\ \frac{3}{2}\sin\theta - 2\cos(2\theta) - 4\sin(2\theta) - 3\cos(3\theta) - 3\sin(3\theta) \\ \qquad\qquad - \cos(4\theta) - 3\sin(4\theta) \\ 12\cos(2\theta) - 54\cos(3\theta) + 18\sin(3\theta) \end{pmatrix}. \quad (7.22)$$

It is easily verified that the function in (7.22) satisfies the solvability condition (1.23). However, the exact solution u of problem (7.1) is not known in this case.

The details mentioned in Remarks 6.11–6.14 apply here as well.

We approximate the solution of (7.1) in this example by means of the MRR method.

7.3.1 Graphical Illustrations

The graphs of the components of $u^{(75)}$ generated by $\psi^{(75)}$ (that is, with 25 points $x^{(k)}$ on ∂S_*) for

$$r \geq 1.01, \quad 0 \leq \theta < 2\pi,$$

together with those of the components of $\psi^{(75)}$, which approximates $u^{\mathscr{A}}|_{\partial S}$, are shown in Fig. 7.11. The plots also show the 25 points $x^{(k)}$ on ∂S_*.

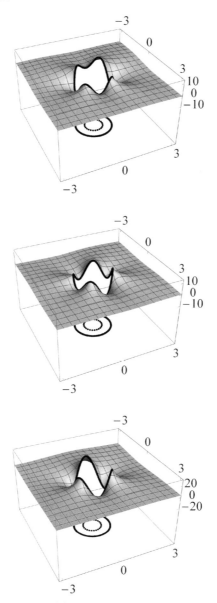

Fig. 7.11 Components of $(u^{\mathscr{A}})^{(75)}$ and $\psi^{(75)}$ (heavy lines around the edge of the hole)

The reason for the restriction $r \geq 1.01$ is that the influence of the singularities of $\big(D(x,y)\big)^{\mathscr{A}}$ and $P(x,y)$ increases considerably for $x \in S^-$ very close to $y \in \partial S$.

Figure 7.12 shows the graphs of the components of $(u^{\mathscr{A}})^{(75)}$ computed from $\psi^{(75)}$ for

$$1.01 \leq r \leq 100, \quad 0 \leq \theta < 2\pi,$$

which illustrate the class $\mathscr{A}$ behavior of the solution away from the boundary. The vertical range of these plots has been truncated to allow better visualization.

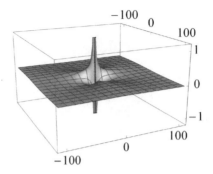

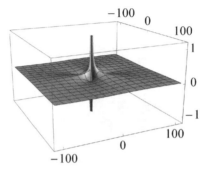

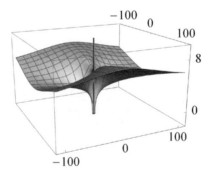

Fig. 7.12 Components of $(u^{\mathscr{A}})^{(75)}$ away from the origin

7.3.2 Computational Procedure

The step-by-step outline of the computational procedure used for solving the exterior Neumann problem starts with the boundary data function $\mathscr{N}$, whose components (in polar coordinates) are graphed in Fig. 7.13.

$\mathscr{N}$ is now digitized on ∂S and the functions $\varphi^{(i)}$ are constructed and digitized as described in Sect. 7.2.2.

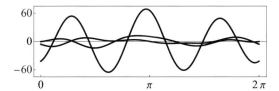

Fig. 7.13 Components of $\mathcal{N}$ in polar coordinates

7.3.3 Computational Results

For the exterior Neumann problem, the approximation $\psi^{(n)}$ of $u^{\mathscr{A}}|_{\partial S}$ is computed by means of the system of equations generated by (7.4) and the $x^{(k)}$. In our example, $\psi^{(75)}$ (see Fig. 7.14) was calculated with 25 points $x^{(k)}$ on ∂S_*. This approximation was then combined with $(u^{\mathscr{A}})^{(75)}$ to generate the graphs in Fig. 7.11, which show good agreement between $(u^{\mathscr{A}})^{(75)}$ and $\psi^{(75)}$, thus producing an indirect validation of our method.

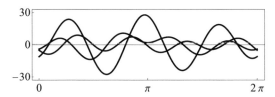

Fig. 7.14 Components of $\psi^{(75)}$ in polar coordinates

7.3.4 Error Analysis

The approximation $\psi^{(n)}$, $n = 3N$, is constructed with N equally spaced points $x^{(k)}$ on ∂S_*. The behavior of the relative error

$$\frac{\| \psi^{(3N)} - u^{\mathscr{A}}|_{\partial S} \|}{\| u^{\mathscr{A}}|_{\partial S} \|}$$

in terms of N determines the efficiency and accuracy of the computational procedure. However, since the exact expression of $u^{\mathscr{A}}|_{\partial S}$ is not known here, an accurate error analysis cannot be performed.

As a substitute, we propose instead an indirect route to obtaining a measure of partial validation of our calculations. We use the approximation $\psi^{(75)}$ of $u^{\mathscr{A}}|_{\partial S}$ as the boundary condition for an exterior Dirichlet problem, and compute the approximate value $(T\bar{u})^{(75)}$ for the solution $\bar{u}$ to the latter by means of the scheme designed in Chap. 6. If our method is sound, then $(T\bar{u})^{(75)}$ should be close to $\mathcal{N} = Tu^{\mathscr{A}}$. Figure 7.15 shows the components of $(T\bar{u})^{(75)}$.

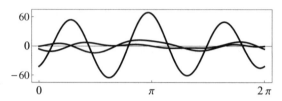

Fig. 7.15 Components of $(T\bar{u})^{(75)}$ in polar coordinates

The components of the error $(T\bar{u})^{(75)}|_{\partial S} - \mathcal{N}$ are displayed in Fig. 7.16.

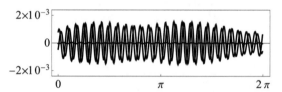

Fig. 7.16 Components of $(T\bar{u})^{(75)} - \mathcal{N}$ in polar coordinates

This plot indicates that the total combined error in this verification exercise is consistent with the error expected when we use MRR to compute $(T\bar{u})^{(75)}$. Repeating our numerical experiment with 200 points $x^{(k)}$ instead of 25 on ∂S_*, we obtained the much smaller relative error

$$\frac{\|(T\bar{u})^{(600)}|_{\partial S} - \mathcal{N}\|}{\|\mathcal{N}\|} = 9.76172 \times 10^{-56}.$$

This confirms the robust nature of the generalized Fourier series method of approximation presented here.

Chapter 8
Exterior Robin Problem

8.1 Computational Algorithm

The exterior Robin problem consists in finding u such that

$$Au(x) = 0, \quad x \in S^-,$$
$$\big(T - \sigma(x)\big)u(x) = \mathcal{R}(x), \quad x \in \partial S, \tag{8.1}$$
$$u \in \mathcal{A}^*,$$

where the symmetric, positive definite, continuous 3×3 matrix function σ and the vector function $\mathcal{R} \in C^{0,\alpha}(\partial S)$, $\alpha \in (0,1)$, are prescribed.

The fact that $u \in \mathcal{A}^*$ means that

$$u = u^{\mathcal{A}} + Fc,$$

where $u^{\mathcal{A}} \in \mathcal{A}$ and Fc is a rigid displacement. Since $Af^{(i)} = 0$ and $Tf^{(i)} = 0$, $i = 1,2,3$, we see that $u^{\mathcal{A}}$ is the solution of the problem

$$Au^{\mathcal{A}} = 0 \quad \text{in } S^-,$$
$$(T - \sigma)u^{\mathcal{A}} = \mathcal{R} + \sigma Fc \quad \text{on } \partial S, \tag{8.2}$$
$$u^{\mathcal{A}} \in \mathcal{A}.$$

Then, by Theorem 1.27, $u^{\mathcal{A}}$ admits the representation formulas

$$u^{\mathcal{A}}(x) = -\int_{\partial S} D(x,y)Tu^{\mathcal{A}}(y)\,ds(y) + \int_{\partial S} P(x,y)u^{\mathcal{A}}(y)\,ds(y), \quad x \in S^-,$$

$$0 = -\int_{\partial S} D(x,y)Tu^{\mathcal{A}}(y)\,ds(y) + \int_{\partial S} P(x,y)u^{\mathcal{A}}(y)\,ds(y), \quad x \in S^+,$$

© Springer Nature Switzerland AG 2020
C. Constanda, D. Doty, *The Generalized Fourier Series Method*, Developments in
Mathematics 65, https://doi.org/10.1007/978-3-030-55849-9_8

which, with $Tu^{\mathscr{A}} = \mathscr{R} + \sigma u^{\mathscr{A}} + \sigma Fc$ on ∂S, become

$$u^{\mathscr{A}}(x) = -\int_{\partial S} D(x,y)(\mathscr{R} + \sigma u^{\mathscr{A}} + \sigma Fc)(y)\,ds(y)$$

$$+\int_{\partial S} P(x,y)u^{\mathscr{A}}(y)\,ds(y), \quad x \in S^-, \qquad (8.3)$$

$$0 = -\int_{\partial S} D(x,y)(\mathscr{R} + \sigma u^{\mathscr{A}} + \sigma Fc)(y)\,ds(y)$$

$$+\int_{\partial S} P(x,y)u^{\mathscr{A}}(y)\,ds(y), \quad x \in S^+. \qquad (8.4)$$

Since $Tu^{\mathscr{A}}$ is the Neumann boundary data of the solution $u^{\mathscr{A}}$ of (8.2), from (1.23) it follows that $\langle Tu^{\mathscr{A}}, f^{(i)}\rangle = 0$, $i = 1,2,3$, translated in our case as

$$\langle \mathscr{R} + \sigma u^{\mathscr{A}} + \sigma Fc, f^{(i)}\rangle = 0, \quad i = 1,2,3.$$

This means that, by Remark 1.24 and what was said in Sect. 1.2.2, D and P can be replaced in (8.3) and (8.4) by $D^{\mathscr{A}}$ and $P^{\mathscr{A}}$, respectively. Recalling that $P = P^{\mathscr{A}}$, we arrive at

$$u^{\mathscr{A}}(x) = -\int_{\partial S} \left(D(x,y)\right)^{\mathscr{A}}(\sigma u^{\mathscr{A}})(y)\,ds(y) + \int_{\partial S} P(x,y)u^{\mathscr{A}}(y)\,ds(y)$$

$$-\int_{\partial S} \left(D(x,y)\right)^{\mathscr{A}}(\mathscr{R} + \sigma Fc)(y)\,ds(y), \quad x \in S^-, \qquad (8.5)$$

$$0 = -\int_{\partial S} \left(D(x,y)\right)^{\mathscr{A}}(\sigma u^{\mathscr{A}})(y)\,ds(y) + \int_{\partial S} P(x,y)u^{\mathscr{A}}(y)\,ds(y)$$

$$-\int_{\partial S} \left(D(x,y)\right)^{\mathscr{A}}(\mathscr{R} + \sigma Fc)(y)\,ds(y), \quad x \in S^+. \qquad (8.6)$$

Equalities (8.5) and (8.6) can also be written for the solution u of (8.1). If we replace $u^{\mathscr{A}} = u - Fc$ in them and take (1.26) into account, a simple reorganization of the terms yields

$$u(x) = \int_{\partial S} \left[P(x,y) - \left(D(x,y)\right)^{\mathscr{A}}\sigma(y)\right]u(y)\,ds(y)$$

$$-\int_{\partial S} \left(D(x,y)\right)^{\mathscr{A}}\mathscr{R}(y)\,ds(y) + (Fc)(x), \quad x \in S^-, \qquad (8.7)$$

$$0 = \int_{\partial S} \left[P(x,y) - \left(D(x,y)\right)^{\mathscr{A}}\sigma(y)\right]u(y)\,ds(y)$$

$$-\int_{\partial S} \left(D(x,y)\right)^{\mathscr{A}}\mathscr{R}(y)\,ds(y) + (Fc)(x), \quad x \in S^+. \qquad (8.8)$$

Let ∂S_* be a simple, closed, C^2-curve lying strictly inside S^+, and let $\{x^{(k)}\}_{k=1}^{\infty}$ be a set of points densely distributed on ∂S_*. We consider the sequence of vector functions on ∂S defined by

$$\mathscr{G} = \{\sigma f^{(i)}, \ i = 1, 2, 3\} \cup \{\varphi^{(jk)}, \ j = 1, 2, 3, \ k = 1, 2, \ldots\},$$

where

$$\varphi^{(jk)}(x) = \left(T - \sigma(x)\right)\left(\left(D(x, x^{(k)})\right)^{\mathscr{A}}\right)^{(j)}. \tag{8.9}$$

8.1 Remark. There are two useful alternative expressions for $\varphi^{(jk)}$, in terms of the rows of matrix $D^{\mathscr{A}}$ and those of matrix P.

(i) By (1.20),

$$\left(D(y, x)\right)^{\mathscr{A}} = \left(\left(D(x, y)\right)^{\mathscr{A}}\right)^{\mathsf{T}},$$

so

$$\left(\left(D(y, x)\right)^{\mathscr{A}}\right)^{(j)} = \left(\left(\left(D(x, y)\right)^{\mathscr{A}}\right)^{\mathsf{T}}\right)^{(j)} = \left(\left(\left(D(x, y)\right)^{\mathscr{A}}\right)_{(j)}\right)^{\mathsf{T}},$$

from which, by (8.9),

$$\varphi^{(jk)}(x) = \left(T - \sigma(x)\right)\left(\left(D(x, x^{(k)})\right)^{\mathscr{A}}\right)^{(j)}$$

$$= \left(T - \sigma(x)\right)\left(\left(\left(D(x^{(k)}, x)\right)^{\mathscr{A}}\right)_{(j)}\right)^{\mathsf{T}}, \quad j = 1, 2, 3, \ k = 1, 2, \ldots. \tag{8.10}$$

(ii) We rewrite (8.10) as

$$\varphi^{(jk)}(x) = T\left(\left(\left(D(x^{(k)}, x)\right)^{\mathscr{A}}\right)_{(j)}\right)^{\mathsf{T}} - \sigma(x)\left(\left(\left(D(x^{(k)}, x)\right)^{\mathscr{A}}\right)_{(j)}\right)^{\mathsf{T}}. \tag{8.11}$$

According to the explanation in Sect. 1.2.2,

$$P(x, y) = \left(P(x, y)\right)^{\mathscr{A}} = \left(T(\partial_y)\left(D(y, x)\right)^{\mathscr{A}}\right)^{\mathsf{T}},$$

so

$$\left(P(x, y)\right)_{(j)} = \left(\left(T(\partial_y)\left(D(y, x)\right)^{\mathscr{A}}\right)^{\mathsf{T}}\right)_{(j)}$$

$$= \left(\left(T(\partial_y)\left(D(y, x)\right)^{\mathscr{A}}\right)^{(j)}\right)^{\mathsf{T}};$$

hence, using this equality and (1.2), we find that

$$\left(\left(P(x, y)\right)_{(j)}\right)^{\mathsf{T}} = \left(T(\partial_y)\left(D(y, x)\right)^{\mathscr{A}}\right)^{(j)}$$

$$= \left(T(\partial_y)\left(\left(D(x, y)\right)^{\mathscr{A}}\right)^{\mathsf{T}}\right)^{(j)}$$

$$= T(\partial_y)\left(\left(\left(D(x, y)\right)^{\mathscr{A}}\right)^{\mathsf{T}}\right)^{(j)}$$

$$= T(\partial_y)\left(\left(\left(D(x, y)\right)^{\mathscr{A}}\right)_{(j)}\right)^{\mathsf{T}}. \tag{8.12}$$

On the other hand,

$$\sigma(x)\left(\left((D(x^{(k)},x))^{\mathscr{A}}\right)_{(j)}\right)^{\mathsf{T}} = (\sigma(x))^{\mathsf{T}}\left(\left((D(x^{(k)},x))^{\mathscr{A}}\right)_{(j)}\right)^{\mathsf{T}}$$

$$= \left(\left((D(x^{(k)},x))^{\mathscr{A}}\right)_{(j)}\sigma(x)\right)^{\mathsf{T}}. \qquad (8.13)$$

Combining (8.12) and (8.13) with (8.11), we arrive at

$$\varphi^{(jk)}(x) = \left((P(x^{(k)},x))_{(j)}\right)^{\mathsf{T}} - \left(\left((D(x^{(k)},x))^{\mathscr{A}}\right)_{(j)}\sigma(x)\right)^{\mathsf{T}}. \qquad (8.14)$$

8.2 Theorem. *The set $\mathscr{G}$ is linearly independent on ∂S and complete in $L^2(\partial S)$.*

Proof. Assuming the opposite, suppose that there are a positive integer N and real numbers c_i and c_{jk}, $i, j = 1, 2, 3$, $k = 1, 2, \ldots$, not all zero, such that

$$\sum_{i=1}^{3} c_i \sigma(x) f^{(i)}(x) + \sum_{j=1}^{3}\sum_{k=1}^{N} c_{jk}\varphi^{(jk)}(x) = 0, \quad x \in \partial S.$$

Setting

$$g(x) = -\sum_{i=1}^{3} c_i f^{(i)}(x) + \sum_{j=1}^{3}\sum_{k=1}^{N} c_{jk}\left((D(x,x^{(k)}))^{\mathscr{A}}\right)^{(j)}$$

and using the fact that the $f^{(i)}$ and the columns of $D^{\mathscr{A}}$ are solutions of the system in (8.1), and that $Tf^{(i)} = 0$, $i = 1, 2, 3$, we see that g satisfies

$$Ag = 0 \quad \text{in } S^-,$$
$$(T - \sigma)g = 0 \quad \text{on } \partial S,$$
$$g \in \mathscr{A}^*;$$

in other words, g is the (unique) regular solution of the homogeneous Robin problem in S^-. According to Theorem 1.29(vi),

$$g = 0 \quad \text{in } \bar{S}^-.$$

Then, by Corollary 3.12, p. 135 in [14],

$$c_i = 0, \quad i = 1, 2, 3,$$

which leads to

$$\sum_{j=1}^{3}\sum_{k=1}^{N} c_{jk}\varphi^{(jk)}(x) = 0, \quad x \in S^-. \qquad (8.15)$$

But

$$\left|\left((D(x,x^{(k)}))^{\mathscr{A}}\right)_{3j}\right| \to \infty \quad \text{as } |x| \to \infty$$

since it contains a $\ln |x|$ term for every $j = 1, 2, 3$ and $k = 1, 2, \ldots, N$. This contradicts (8.15), so all $c_{jk} = 0$, proving the linear independence of $\mathcal{G}$.

Suppose now that a function $q \in L^2(\partial S)$ satisfies

$$\langle q, \sigma f^{(i)} \rangle = \int_{\partial S} (\sigma f^{(i)})^{\mathsf{T}} q \, ds = 0, \quad i = 1, 2, 3, \tag{8.16}$$

$$\langle q, \varphi^{(jk)} \rangle = \int_{\partial S} (\varphi^{(jk)})^{\mathsf{T}} q \, ds = 0, \quad j = 1, 2, 3, \ k = 1, 2, \ldots . \tag{8.17}$$

According to (5.9) with v and w replaced by q and $f^{(i)}$, respectively, and (8.16),

$$\langle \sigma q, f^{(i)} \rangle = \langle q, \sigma f^{(i)} \rangle = 0; \tag{8.18}$$

hence, by Remark 1.24,

$$\int_{\partial S} D(x, y)(\sigma q)(y) \, ds(y) = \int_{\partial S} (D(x, y))^{\mathscr{A}} (\sigma q)(y) \, ds(y).$$

Next, since $P = P^{\mathscr{A}}$, the double-layer potential with L^2-density q can be written as

$$(Wq)(x) = \int_{\partial S} P(x, y) q(y) \, ds(y) = \int_{ds} (P(x, y))^{\mathscr{A}} q(y) \, ds(y)$$

$$= \int_{\partial S} \left(T(\partial_y)(D(y, x))^{\mathscr{A}} \right)^{\mathsf{T}} q(y) \, ds(y), \quad x \in S^+ \cup S^-.$$

Similarly, by Theorem 1.32(ii), the single-layer potential with L^2-density σq has the expression

$$(V(\sigma q))(x) = (V(\sigma q))^{\mathscr{A}}(x)$$

$$= \int_{\partial S} (D(x, y))^{\mathscr{A}} \sigma(y) q(y) \, ds(y)$$

$$= \int_{\partial S} \left((D(y, x))^{\mathscr{A}} \right)^{\mathsf{T}} \sigma(y) q(y) \, ds(y)$$

$$= \int_{\partial S} \left(\sigma^{\mathsf{T}}(y) \left((D(y, x))^{\mathscr{A}} \right) \right)^{\mathsf{T}} q(y) \, ds(y)$$

$$= \int_{\partial S} \left(\sigma(y) \left((D(y, x))^{\mathscr{A}} \right) \right)^{\mathsf{T}} q(y) \, ds(y), \quad x \in S^+ \cup S^-.$$

Then

$$(Wq)(x) - (V(\sigma q))(x)$$
$$= \int_{\partial S} \left\{ \left(T(\partial_y)(D(y,x))^{\mathscr{A}} \right)^{\mathsf{T}} - \left(\sigma(y)(D(y,x))^{\mathscr{A}} \right)^{\mathsf{T}} \right\} q(y) \, ds(y)$$
$$= \int_{\partial S} \left((T(\partial_y) - \sigma(y))(D(y,x))^{\mathscr{A}} \right)^{\mathsf{T}} q(y) \, ds(y), \quad x \in S^+ \cup S^-,$$

so

$$(Wq)(x^{(k)}) - (V(\sigma q))(x^{(k)}) = \int_{\partial S} \left((T - \sigma(x))(D(x,x^{(k)}))^{\mathscr{A}} \right)^{\mathsf{T}} q(x) \, ds(x),$$

from which, by (8.17),

$$\left((Wq)(x^{(k)}) \right)_j - \left((V(\sigma q))(x^{(k)}) \right)_j$$
$$= \int_{\partial S} \left(\left((T - \sigma(x))(D(x,x^{(k)}))^{\mathscr{A}} \right)^{\mathsf{T}} \right)_{(j)} q(x) \, ds(x)$$
$$= \int_{\partial S} \left(\left((T - \sigma(x))(D(x,x^{(k)}))^{\mathscr{A}} \right)^{(j)} \right)^{\mathsf{T}} q(x) \, ds(x)$$
$$= \int_{\partial S} (\varphi^{(jk)})^{\mathsf{T}} q \, ds = \langle \varphi^{(jk)}, q \rangle = 0.$$

Since the set $\{x^{(k)}\}_{k=1}^{\infty}$ is dense on ∂S_*, we conclude that

$$(Wq)(x) - (V(\sigma q))(x) = 0, \quad x \in \partial S_*. \tag{8.19}$$

Also, by Theorems 4.18, p. 99 and 4.2, p. 84 in [15], (8.18), and (8.19), the function

$$U(x) = (Wq)(x) - (V(\sigma q))(x), \quad x \in S^+ \cup S^-,$$

is the (unique) solution of the interior Dirichlet problem

$$AU = 0 \quad \text{in } S_*^+,$$
$$U = 0 \quad \text{on } \partial S_*;$$

hence,

$$U = 0 \quad \text{in } \bar{S}_*^+$$

and, by analyticity,

$$U = 0 \quad \text{in } S^+.$$

This implies that

$$U^+ = 0 \quad \text{on } \partial S, \tag{8.20}$$

$$(TU)^+ = 0 \quad \text{on } \partial S, \tag{8.21}$$

which, by Theorems 4.20, p.99 and 4.21, p. 100 in [15], leads to

$$-\frac{1}{2}q + W_0 q - V(\sigma q) = 0 \quad \text{on } \partial S,$$

or, since $V(\sigma q) = V^{\mathscr{A}}(\sigma q)$,

$$-\frac{1}{2}q(x) + \int_{\partial S} P(x,y)q(y)\,ds(y) - \int_{\partial S} D(x,y)\sigma(y)q(y)\,ds(y) = 0, \quad x \in \partial S.$$

This equation is of the form

$$-\frac{1}{2}q(x) + \int_{\partial S} k(x,y)q(y)\,ds(y) = 0, \quad x \in \partial S,$$

where, according to the properties of the layer potentials investigated in Section 4.2, p. 94 in [15],

$$k(x,y) = P(x,y) - D(x,y)\sigma(y)$$

is a so-called α-regular singular kernel on ∂S. Then, repeating the proof of Theorem 6.12, p. 145 in [15], we conclude that $q \in C^{1,\alpha}(\partial S)$ for any $\alpha \in (0,1)$, and we can use the properties of V and W given by Theorem 1.33 to compute

$$
\begin{aligned}
(TU)^- &= \left(T\big(Wq - V(\sigma q)\big)\right)^- = \left(T(Wq)\right)^- - \left(TV(\sigma q)\right)^- \\
&= \left(T(Wq)\right)^- + \frac{1}{2}\sigma q - W_0^*(\sigma q) \\
&= \left(T(Wq)\right)^+ + \frac{1}{2}\sigma q - \left[-\frac{1}{2}\sigma q + \left(TV(\sigma q)\right)^+\right] \\
&= \left(T(Wq)\right)^+ - \left(TV(\sigma q)\right)^+ + \sigma q \\
&= \left(T\big(Wq - V(\sigma q)\big)\right)^+ + \sigma q = (TU)^+ + \sigma q = \sigma q \tag{8.22}
\end{aligned}
$$

and

$$
\begin{aligned}
(\sigma U)^- &= \sigma\left[(Wq)^- - \big(V(\sigma q)\big)^-\right] \\
&= \sigma\left[\frac{1}{2}q + W_0 q - \big(V(\sigma q)\big)^-\right] \\
&= \sigma\left[\frac{1}{2}q + \left(\frac{1}{2}q + (Wq)^+\right) - \big(V(\sigma q)\big)^+\right] \\
&= \sigma(q + U^+) = \sigma q + \sigma U^+ = \sigma q,
\end{aligned}
$$

so, by (8.20) and (8.21),

$$\big((T-\sigma)U\big)^- = 0 \quad \text{on } \partial S.$$

Therefore, U is the (unique) regular solution of the homogeneous Robin problem

$$AU = 0 \quad \text{in } S^-,$$

$$\left((T - \sigma)U\right)^- = 0 \quad \text{on } \partial S,$$

$$U \in \mathscr{A},$$

which, by Theorem 1.29(vi), means that

$$U = 0 \quad \text{in } S^+.$$

Then, taking (8.21) into account, we find that

$$(TU)^- = 0 \quad \text{on } \partial S,$$

and that, by (8.22),

$$\sigma q = 0 \quad \text{on } \partial S.$$

Since σ is positive definite, we conclude that $q = 0$, thus proving that, by Theorem 3.1, $\mathscr{G}$ is a complete set in $L^2(\partial S)$. □

We order the elements of $\mathscr{G}$ as the sequence

$$\sigma f^{(1)}, \sigma f^{(2)}, \sigma f^{(3)}, \varphi^{(11)}, \varphi^{(21)}, \varphi^{(31)}, \varphi^{(12)}, \varphi^{(22)}, \varphi^{(32)}, \ldots$$

and re-index them, noticing that for each $i = 4, 5, \ldots$, there is a unique pair $\{j, k\}$, $j = 1, 2, 3$, $k = 1, 2, \ldots$, such that $i = j + 3k$. Then

$$\mathscr{G} = \{\varphi^{(1)}, \varphi^{(2)}, \varphi^{(3)}, \varphi^{(4)}, \varphi^{(5)}, \varphi^{(6)}, \varphi^{(7)}, \varphi^{(8)}, \varphi^{(9)}, \ldots\}, \tag{8.23}$$

where

$$\varphi^{(i)} = \sigma f^{(i)}, \quad i = 1, 2, 3, \tag{8.24}$$

and

$$\varphi^{(i)} = \varphi^{(jk)}, \quad j = 1, 2, 3, \ k = 1, 2, \ldots, \ i = j + 3k = 4, 5, \ldots. \tag{8.25}$$

8.1.1 Row Reduction Method

Our intention is to construct a nonsingular linear system of finitely many algebraic equations to approximate the solution u of (8.1).

Rewriting the representation formulas (8.7) and (8.8) as

$$u(x) = \langle P(x, \cdot) - (D(x, \cdot))^{\mathscr{A}} \sigma, u \rangle$$
$$- \langle (D(x, \cdot))^{\mathscr{A}}, \mathscr{R} \rangle + (Fc)(x), \quad x \in S^-, \tag{8.26}$$

$$\langle P(x,\cdot) - (D(x,\cdot))^{\mathscr{A}}\sigma, u\rangle + (Fc)(x)$$
$$= \langle (D(x,\cdot))^{\mathscr{A}}, \mathscr{R}\rangle, \quad x \in S^+, \tag{8.27}$$

we would normally use (8.27) to approximate $u|_{\partial S}$, and then approximate u in S^- by means of (8.26). However, this operation cannot be performed since (8.27) contains the rigid displacement

$$Fc = \sum_{i=1}^{3} c_i f^{(i)},$$

which may not be known a priori and needs to be determined as part of the computation process.

We set

$$\psi = u|_{\partial S}.$$

Since the $x^{(k)}$ are points in S^+, from (8.27) it follows that

$$\langle P(x^{(k)},\cdot) - (D(x^{(k)},\cdot))^{\mathscr{A}}\sigma, \psi\rangle + (Fc)(x^{(k)})$$
$$= \langle (D(x^{(k)},\cdot))^{\mathscr{A}}, \mathscr{R}\rangle, \quad j = 1,2,3, \ k = 1,2,\ldots,$$

from which

$$\left\langle \left((P(x^{(k)},\cdot))_{(j)}\right)^{\mathsf{T}} - \left(\left((D(x^{(k)},\cdot))^{\mathscr{A}}\sigma\right)_{(j)}\right)^{\mathsf{T}}, \psi\right\rangle + (Fc)_j(x^{(k)})$$
$$= \left\langle \left(\left((D(x^{(k)},\cdot))^{\mathscr{A}}\right)_{(j)}\right)^{\mathsf{T}}, \mathscr{R}\right\rangle, \quad j = 1,2,3, \ k = 1,2,\ldots. \tag{8.28}$$

We have

$$\langle (D(x,\cdot))^{\mathscr{A}}\sigma, \psi\rangle = \int_{\partial S} (D(x,y))^{\mathscr{A}}\sigma(y)\psi(y)\,ds(y) = \langle (D(x,\cdot))^{\mathscr{A}}, \sigma\psi\rangle,$$

so, by (5.9),

$$\left\langle \left(\left((D(x^{(k)},\cdot))^{\mathscr{A}}\sigma\right)_{(j)}\right)^{\mathsf{T}}, \psi\right\rangle = \left\langle \left(\left((D(x^{(k)},\cdot))^{\mathscr{A}}\right)_{(j)}\right)^{\mathsf{T}}, \sigma\psi\right\rangle$$
$$= \left\langle \sigma\left(\left((D(x^{(k)},\cdot))^{\mathscr{A}}\right)_{(j)}\right)^{\mathsf{T}}, \psi\right\rangle$$
$$= \left\langle \left(\left((D(x^{(k)},\cdot))^{\mathscr{A}}\right)_{(j)}\sigma^{\mathsf{T}}\right)^{\mathsf{T}}, \psi\right\rangle$$
$$= \left\langle \left(\left((D(x^{(k)},\cdot))^{\mathscr{A}}\right)_{(j)}\sigma\right)^{\mathsf{T}}, \psi\right\rangle.$$

Replacing in (8.28) and using (8.14) and (8.25), we deduce that

$$\langle \psi, \varphi^{(i)}\rangle = \langle \psi, \varphi^{(jk)}\rangle$$
$$= \left\langle \left((P(x^{(k)},\cdot))_{(j)}\right)^{\mathsf{T}} - \left(\left((D(x^{(k)},\cdot))^{\mathscr{A}}\right)_{(j)}\sigma\right)^{\mathsf{T}}, \psi\right\rangle$$

$$= \left\langle \left(\left(P(x^{(k)},\cdot) \right)_{(j)} \right)^{\mathsf{T}} - \left(\left(\left(D(x^{(k)},\cdot) \right)^{\mathscr{A}} \sigma \right)_{(j)} \right)^{\mathsf{T}}, \psi \right\rangle$$

$$= \left\langle \left(\left(\left(D(x,\cdot) \right)^{\mathscr{A}} \right)_{(j)} \right)^{\mathsf{T}}, \mathscr{R} \right\rangle - (Fc)_j(x^{(k)}),$$

$$j = 1,2,3, \quad k = 1,2,\ldots, \quad i = j+3k = 4,5,\ldots. \tag{8.29}$$

At the same time, since $Tu = \mathscr{R} + \sigma\psi$ is the Neumann boundary data of the solution, we have

$$\langle Tu, f^{(i)} \rangle = 0, \quad i = 1,2,3,$$

so

$$\langle \sigma\psi, f^{(i)} \rangle = -\langle \mathscr{R}, f^{(i)} \rangle, \quad i = 1,2,3,$$

which, by (5.9), yields

$$\langle \psi, \sigma f^{(i)} \rangle = \langle \sigma\psi, f^{(i)} \rangle = -\langle \mathscr{R}, f^{(i)} \rangle, \quad i = 1,2,3. \tag{8.30}$$

In view of Theorem 8.2, there is a unique expansion

$$\psi = \sum_{h=1}^{\infty} c_h \varphi^{(h)},$$

so we can seek an approximation for ψ of the form

$$\psi^{(n)} = (u|_{\partial S})^{(n)} = \sum_{h=1}^{n} c_h \varphi^{(h)}. \tag{8.31}$$

8.3 Remark. For symmetry and ease of computation, in the construction of the approximation $\psi^{(n)}$ we use the subsequence with $n = 3N + 3$, where N is the number of points $x^{(k)}$ chosen on ∂S_*. This choice inserts in the computation, for each value of $k = 1,2,\ldots,N$, all three columns/rows of $D^{\mathscr{A}}$ and P involved in the definition (8.9) and its alternatives (8.10) and (8.14) of the vector functions $\varphi^{(i)}$, $i = 4,5,\ldots,n = 3N+3$, with the re-indexing (8.25).

The combinations of (8.31) and (8.30), and then of (8.31) and (8.29), yield the approximate equalities

$$\sum_{h=1}^{n} c_h \langle \sigma f^{(i)}, \varphi^{(h)} \rangle = -\langle f^{(i)}, \mathscr{R} \rangle, \quad i = 1,2,3, \tag{8.32}$$

$$\sum_{h=1}^{n} c_h \langle \varphi^{(i)}, \varphi^{(h)} \rangle + \sum_{l=1}^{3} c_l f_j^{(l)}(x^{(k)}) = \left\langle \left(\left(\left(D(x^{(k)},\cdot) \right)^{\mathscr{A}} \right)_{(j)} \right)^{\mathsf{T}}, \mathscr{R} \right\rangle,$$

$$i = j+3k = 4,5,\ldots,3N+3. \tag{8.33}$$

Equations (8.32) and (8.33) form our system for determining the $3N+3$ coefficients c_i. In expanded form, this system is

$$
\begin{pmatrix}
\langle \sigma f^{(1)}, \varphi^{(1)} \rangle & \langle \sigma f^{(1)}, \varphi^{(2)} \rangle & \cdots & \langle \sigma f^{(1)}, \varphi^{(n)} \rangle \\
\langle \sigma f^{(2)}, \varphi^{(1)} \rangle & \langle \sigma f^{(2)}, \varphi^{(2)} \rangle & \cdots & \langle \sigma f^{(2)}, \varphi^{(n)} \rangle \\
\langle \sigma f^{(3)}, \varphi^{(1)} \rangle & \langle \sigma f^{(3)}, \varphi^{(2)} \rangle & \cdots & \langle \sigma f^{(3)}, \varphi^{(n)} \rangle \\
\langle \varphi^{(4)}, \varphi^{(1)} \rangle & \langle \varphi^{(4)}, \varphi^{(2)} \rangle & \cdots & \langle \varphi^{(4)}, \varphi^{(n)} \rangle \\
\langle \varphi^{(5)}, \varphi^{(1)} \rangle & \langle \varphi^{(5)}, \varphi^{(2)} \rangle & \cdots & \langle \varphi^{(5)}, \varphi^{(n)} \rangle \\
\vdots & \vdots & & \vdots \\
\langle \varphi^{(n)}, \varphi^{(1)} \rangle & \langle \varphi^{(n)}, \varphi^{(2)} \rangle & \cdots & \langle \varphi^{(n)}, \varphi^{(n)} \rangle
\end{pmatrix}
\begin{pmatrix}
c_1 \\ c_2 \\ c_3 \\ c_4 \\ c_5 \\ \vdots \\ c_n
\end{pmatrix}
$$

$$
+
\begin{pmatrix}
0 & 0 & 0\,0\,0\,0\dots0 \\
0 & 0 & 0\,0\,0\,0\dots0 \\
0 & 0 & 0\,0\,0\,0\dots0 \\
1 & 0 & 0\,0\,0\,0\dots0 \\
0 & 1 & 0\,0\,0\,0\dots0 \\
-x_1^{(1)} & -x_2^{(1)} & 1\,0\,0\,0\dots0 \\
\vdots & \vdots & \vdots\,\vdots\,\vdots\,\vdots \quad\vdots \\
-x_1^{(N)} & -x_2^{(N)} & 1\,0\,0\,0\dots0
\end{pmatrix}
\begin{pmatrix}
c_1 \\ c_2 \\ c_3 \\ c_4 \\ c_5 \\ c_6 \\ \vdots \\ c_n
\end{pmatrix}
$$

$$
=
\begin{pmatrix}
-\langle f^{(1)}, \mathscr{R} \rangle \\
-\langle f^{(2)}, \mathscr{R} \rangle \\
-\langle f^{(3)}, \mathscr{R} \rangle \\
\langle \big((D(x^{(1)},\cdot))_{(1)} \big)^{\mathsf{T}}, \mathscr{R} \rangle \\
\langle \big((D(x^{(1)},\cdot))_{(2)} \big)^{\mathsf{T}}, \mathscr{R} \rangle \\
\vdots \\
\langle \big((D(x^{(N)},\cdot))_{(3)} \big)^{\mathsf{T}}, \mathscr{R} \rangle
\end{pmatrix}.
\tag{8.34}
$$

System (8.34) is nonsingular, so it has a unique solution that generates the approximation $\psi^{(n)}$ in (8.31) and the approximation $Fc^{(n)}$ of the rigid displacement.

8.4 Remark. We have deliberately written $\sigma f^{(i)}$ instead of $\varphi^{(i)}$ in the inner products in the first three rows in (8.34), to emphasize that those equations have a different source than the rest.

Using (8.26), we now construct the vector functions

$$
u^{(n)}(x) = \langle P(x,\cdot) - (D(x,\cdot))^{\mathscr{A}} \sigma, \psi^{(n)} \rangle
$$
$$
- \langle (D(x,\cdot))^{\mathscr{A}}, \mathscr{R} \rangle + (Fc)^{(n)}(x), \quad x \in S^-,
\tag{8.35}
$$

and

$$(u^{\mathscr{A}})^{(n)}(x) = u^{(n)}(x) - (Fc)^{(n)}(x)$$
$$= \langle P(x,\cdot) - (D(x,\cdot))^{\mathscr{A}}\sigma, \psi^{(n)}\rangle - \langle (D(x,\cdot))^{\mathscr{A}}, \mathscr{R}\rangle, \quad x \in S^-. \quad (8.36)$$

8.5 Remark. In the interior Dirichlet problem, the rigid displacement in the solution is computed separately. In the interior Neumann problem, it remains completely arbitrary and plays no role in the calculation. In both these cases, the linear algebraic system generated by the boundary data has a nonzero null space that needs special attention. Here, by contrast, the rigid displacement cannot be isolated from the full numerical process. This helps simplify the formulation of the algorithm and the computation.

8.6 Remark. In [15], Section 6.2, p. 141, it is shown that the exact rigid displacement Fc can be computed directly from $\mathscr{R}$ with the help of the (unique) solution of a certain boundary integral equation on ∂S. In view of this, we may redefine $u^{(n)}$ alternatively as

$$u^{(n)}(x) = \langle P(x,\cdot) - (D(x,\cdot))^{\mathscr{A}}\sigma, \psi^{(n)}\rangle$$
$$- \langle (D(x,\cdot))^{\mathscr{A}}, \mathscr{R}\rangle + (Fc)(x), \quad x \in S^-. \quad (8.37)$$

This form is preferred in analytic considerations, whereas (8.35) is the version of choice in computational work.

8.7 Theorem. *The vector function $u^{(n)}$ defined by (8.37) is an approximation of the solution u of problem (8.1) in the sense that $u^{(n)} \to u$ uniformly on any closed and bounded subdomain S' of S^-.*

Proof. By (8.26) and (8.37),

$$u(x) - u^{(n)}(x) = \langle P(x,\cdot) - (D(x,\cdot))^{\mathscr{A}}\sigma, \psi\rangle$$
$$- \langle P(x,\cdot) - (D(x,\cdot))^{\mathscr{A}}\sigma, \psi^{(n)}\rangle$$
$$= \langle P(x,\cdot) - (D(x,\cdot))^{\mathscr{A}}\sigma, \psi - \psi^{(n)}\rangle, \quad x \in S^-.$$

Since

$$\lim_{n \to \infty} \|\psi - \psi^{(n)}\| = 0,$$

the argument now proceeds as in the proof of Theorem 5.7 in Sect. 5.2, with D replaced by $P - D^{\mathscr{A}}\sigma$. $\quad\square$

8.1.2 Orthonormalization Method

After the re-indexing (8.23)–(8.25), we orthonormalize the set $\{\varphi^{(i)}\}_{i=1}^n$ as

$$\{\omega^{(i)}\}_{i=1}^n, \quad n = 3N + 3.$$

Ordinarily, we would then try to compute the approximation $\psi^{(n)}$ of ψ by means of formula (2.2); that is,

$$
\psi^{(n)} = \sum_{i=1}^{n} \langle \psi, \omega^{(i)} \rangle \omega^{(i)}
$$

$$
= \sum_{i=1}^{n} \left\langle \psi, \sum_{j=1}^{i} \bar{r}_{ij} \varphi^{(j)} \right\rangle \left(\sum_{h=1}^{i} \bar{r}_{ih} \varphi^{(h)} \right)
$$

$$
= \sum_{i=1}^{n} \sum_{j=1}^{i} \bar{r}_{ij} \langle \psi, \varphi^{(j)} \rangle \left(\sum_{h=1}^{i} \bar{r}_{ih} \varphi^{(h)} \right) \quad \text{on } \partial S.
$$

The coefficients $\langle \psi, \varphi^{(i)} \rangle$, $i = 1, 2, 3$, are given by (8.30). For $i = 4, 5, \ldots, 3N + 3$, they should normally be computed from (8.29). However, that formula requires knowledge of the rigid displacement Fc, which we can only determine a priori by the procedure described in Remark 8.6. Since this takes us outside our computational algorithm, we prefer to solve the exterior Robin problem by means of the MRR method.

8.2 Numerical Example 1

Let S^+ be the disk of radius 1 centered at origin, let (after rescaling and non-dimensionalization)

$$
h = 0.5, \quad \lambda = \mu = 1,
$$

and let

$$
\sigma(x) = \begin{pmatrix} 11 & 2 & 1 \\ 2 & 8 & -2 \\ 1 & -2 & 11 \end{pmatrix}.
$$

We choose the boundary condition function $\mathcal{R}$ (in polar coordinates on ∂S) in (8.1) to be

$$
\mathcal{R}(x_p) = \begin{pmatrix} -33 - 8\sin\theta - 23\cos(2\theta) - 12\sin(2\theta) - 3\cos(3\theta) \\ \quad + 63\sin(3\theta) + 13\cos(4\theta) + 46\sin(4\theta) \\ -30 - 11\cos\theta - 6\sin\theta - 16\cos(2\theta) - 17\sin(2\theta) - 48\cos(3\theta) \\ \quad - 18\sin(3\theta) - 26\cos(4\theta) + 23\sin(4\theta) \\ 45 + 24\cos\theta + 21\sin\theta + 2\cos(2\theta) + 58\sin(2\theta) + 69\cos(3\theta) \\ \quad + 123\sin(3\theta) + 10\cos(4\theta) \end{pmatrix}.
$$

It is easily verified that the exact solution u of (8.1) in this case is

$$u(x) = \begin{pmatrix} 3 + (x_1^2 + x_2^2)^{-1}x_2 \\ \quad + (x_1^2 + x_2^2)^{-2}(2x_1^2 + 4x_1x_2 - 2x_2^2 - 3x_1^2x_2 + x_2^3) \\ \quad + (x_1^2 + x_2^2)^{-3}(-12x_1^2x_2 + 4x_2^3 - 2x_1^4 - 16x_1^3x_2 \\ \qquad\qquad + 12x_1^2x_2^2 + 16x_1x_2^3 - 2x_2^4) \\[4pt] 2 + (x_1^2 + x_2^2)^{-1}x_1 \\ \quad + (x_1^2 + x_2^2)^{-2}(2x_1^2 - 2x_2^2 + x_1^3 - 3x_1x_2^2) \\ \quad + (x_1^2 + x_2^2)^{-3}(4x_1^3 - 12x_1x_2^2 + 4x_1^4 - 8x_1^3x_2 \\ \qquad\qquad - 24x_1^2x_2^2 + 8x_1x_2^3 + 4x_2^4) \\[4pt] -4 - 3x_1 - 2x_2 + (x_1^2 + x_2^2)^{-1}(x_1 - 2x_1x_2) \\ \quad + (x_1^2 + x_2^2)^{-2}(-10x_1x_2 - x_1^3 - 6x_1^2x_2 + 3x_1x_2^2 + 2x_2^3) \\ \quad + (x_1^2 + x_2^2)^{-3}(-6x_1^3 - 36x_1^2x_2 + 18x_1x_2^2 + 12x_2^3) \end{pmatrix},$$

or, in polar coordinates,

$$u(x_p) = \begin{pmatrix} 3 + r^{-1}[\sin\theta - \sin(3\theta)] \\ \quad + r^{-2}[2\cos(2\theta) + 2\sin(2\theta) - 2\cos(4\theta) - 4\sin(4\theta)] \\ \quad + r^{-3}[-4\sin(3\theta)] \\[4pt] 2 + r^{-1}[\cos\theta + \cos(3\theta)] \\ \quad + r^{-2}[2\cos(2\theta) + 4\cos(4\theta) - 2\sin(4\theta)] \\ \quad + r^{-3}[4\cos(3\theta)] \\[4pt] r(-3\cos\theta - 2\sin\theta) - 4 - \sin(2\theta) \\ \quad + r^{-1}[\cos\theta - \cos(3\theta) - 2\sin(3\theta)] \\ \quad + r^{-2}[-5\sin(2\theta)] \\ \quad + r^{-3}[-6\cos(3\theta) - 12\sin(3\theta)] \end{pmatrix}. \qquad (8.38)$$

8.8 Remark. Using (8.38) and (1.15), we easily convince ourselves that the solution belongs to class $\mathscr{A}^*$, with the class $\mathscr{A}$ component $u^{\mathscr{A}}$ defined by the coefficients

$$\begin{aligned} m_0 &= 1, \\ m_1 &= 0, & m_2 &= 0, \\ m_3 &= 2, & m_4 &= -2, \\ m_5 &= 2, & m_6 &= -1, \\ m_7 &= -2, & m_8 &= 0, \\ m_9 &= -2, & m_{10} &= 0, \\ m_{11} &= -5, & m_{12} &= 0. \end{aligned}$$

The procedure described in Remark 1.11 also shows that this solution contains the rigid displacement $f = 3f^{(1)} + 2f^{(2)} - 4f^{(3)}$, alternatively written as

$$f = Fc, \quad c = (3, 2, -4)^{\mathsf{T}}.$$

We select ∂S_* to be the circle of radius $1/2$ centered at the origin.

The details mentioned in Remarks 6.11–6.14 apply here as well.

As already mentioned, we approximate the solution of problem (8.1) in this example by means of the MRR method.

8.2.1 Graphical Illustrations

The graphs of the components of $(u^{\mathscr{A}})^{(63)}$ computed from $\psi^{(63)}$ (that is, with 20 points $x^{(k)}$ on ∂S_*) for

$$r \geq 1.01, \quad 0 \leq \theta < 2\pi,$$

together with those of the components of $\psi^{(63)} - f$, which approximates $u^{\mathscr{A}}|_{\partial S}$, are shown in Fig. 8.1.

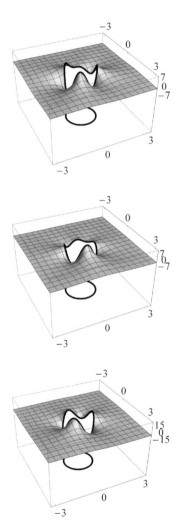

Fig. 8.1 Components of $(u^{\mathscr{A}})^{(63)}$ and $\psi^{(63)} - f$ (heavy lines around the edge of the hole)

The reason for the restriction $r \geq 1.01$ is the increasing influence of the singularities of $\left(D(x,y)\right)^{\mathscr{A}}$ and $P(x,y)$ for $x \in S^-$ very close to $y \in \partial S$.

Figure 8.2 shows the graphs of the components of $(u^{\mathscr{A}})^{(63)}$ computed from $\psi^{(63)}$ for

$$1.01 \leq r \leq 100, \quad 0 \leq \theta < 2\pi,$$

illustrating the class $\mathscr{A}$ behavior of the solution away from the boundary. The vertical range of these plots has been truncated to allow better visualization.

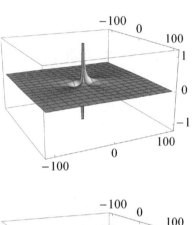

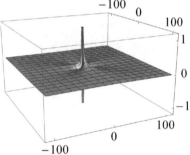

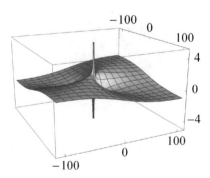

Fig. 8.2 Components of $(u^{\mathscr{A}})^{(63)}$ away from the boundary

The graphs of the components of the actual error $(u^{\mathscr{A}})^{(63)} - u^{\mathscr{A}}$ together with ∂S and the 20 points $x^{(k)}$ on ∂S_* can be seen in Fig. 8.3. The approximation is 4–5 digits of accuracy near ∂S and improves significantly away from the boundary.

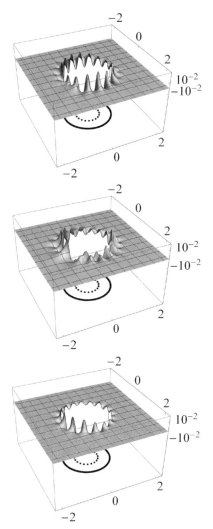

Fig. 8.3 Components of the error $(u^{\mathscr{A}})^{(63)} - u^{\mathscr{A}}$

8.2.2 Computational Procedure

The step-by-step outline of the computational procedure used for solving the exterior Robin problem starts with the boundary data function $\mathscr{R}$, whose components (in polar coordinates) are graphed in Fig. 8.4.

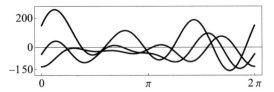

Fig. 8.4 Components of $\mathscr{R}$ in polar coordinates

$\mathscr{R}$ is digitized on ∂S with a sufficient number of points to guarantee that the accuracy of all quadrature calculations significantly exceeds the expected magnitude of the error produced by our approximation method. This allows the method error to be isolated from any quadrature errors. For this example, 100 discrete points on ∂S prove adequate to reach our objective. Next, the functions $\varphi^{(i)}$ are constructed as described in (8.24) and (8.25), and are digitized as well. If instead of MRR we were to select an orthonormalization method—say, MGS—with the rigid displacement computed a priori as explained in Remark 8.6, then the $\omega^{(i)}$ would be digitized in the same way.

The graphs of typical three-component vector functions $\varphi^{(i)}$ and $\omega^{(i)}$ are displayed in Figs. 8.5 and 8.6, respectively. The increasing number of oscillations in the latter eventually produces computational difficulties in the evaluation of the inner products occurring in the construction of additional orthonormal vector functions.

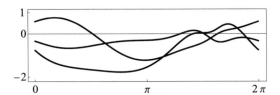

Fig. 8.5 Typical vector functions $\varphi^{(i)}$ in polar coordinates

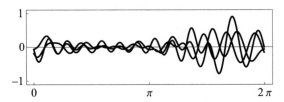

Fig. 8.6 Typical vector functions $\omega^{(i)}$ in polar coordinates

8.2.3 Computational Results

For the exterior Robin problem, the approximation $\psi^{(n)}$ of $u|_{\partial S}$ is computed by means of the system of equations generated by (8.26), (8.27), (8.30), and the $x^{(k)}$.

In our example, $\psi^{(63)}$ was calculated with 20 points $x^{(k)}$ on ∂S_* (see Fig. 8.7). As Fig. 8.8 shows, this approximation is quite good when compared with the exact result for $u|_{\partial S}$ derived from the known solution u.

Equation (8.35) and $\psi^{(63)}$ generate the graphs displayed in Figs. 8.1–8.3.

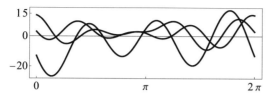

Fig. 8.7 Components of $\psi^{(63)}$ in polar coordinates

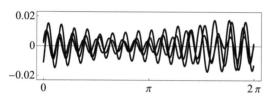

Fig. 8.8 Components of the error $\psi^{(63)} - u|_{\partial S}$ in polar coordinates

8.2.4 Error Analysis

The approximation $\psi^{(n)}$, $n = 3N + 3$, is computed using N equally spaced points $x^{(k)}$ on ∂S_*. The behavior of the relative error

$$\frac{\|\psi^{(3N+3)} - u|_{\partial S}\|}{\|u|_{\partial S}\|}$$

as a function of N determines the efficiency and accuracy of the computational procedure. The logarithmic plot in Fig. 8.9 illustrates the size of the relative error in terms of N.

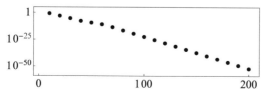

Fig. 8.9 The relative error $\|\psi^{(3N+3)} - u|_{\partial S}\|/\|u|_{\partial S}\|$ for $N = 5, 10, \ldots, 200$

   This plot strongly suggests that the relative error improves exponentially as N increases. Fitting a linear curve to the logarithmic data for $N > 50$ gives rise to the model

$$6.94863 - 0.2953N.$$

A comparison of the logarithmic data and the linear model is shown in Fig. 8.10.

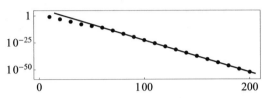

Fig. 8.10 The relative error $\|\psi^{(3N+3)} - u|_{\partial S}\|/\|u|_{\partial S}\|$ and the linear model for $N = 5, 10, \ldots, 200$

   The two are seen to be in good agreement for $N > 50$, so the relative error may be modeled by

$$\frac{\|\psi^{(3N+3)} - u|_{\partial S}\|}{\|u|_{\partial S}\|} = 8.88452 \times 10^6 \times 10^{-0.2953N}$$

$$= 8.88452 \times 10^6 \times 0.506641^N.$$

   We remark that system (8.34) also generates the approximation

$$f^{(n)} = c_1^{(n)} f^{(1)} + c_2^{(n)} f^{(2)} + c_3^{(n)} f^{(3)},$$

or

$$f^{(n)} = Fc^{(n)}, \quad c^{(n)} = (c_1^{(n)}, c_2^{(n)}, c_3^{(n)})^\mathsf{T},$$

of the rigid displacement $f = Fc = 3f^{(1)} + 2f^{(2)} - 4f^{(3)}$. In our case, this is

$$f^{(63)} = 2.99999999045848 \, f^{(1)} + 2.00000000573438 \, f^{(2)}$$

$$- 4.0000000016693 \, f^{(3)},$$

with error

$$f^{(63)} - f = -9.54152 \times 10^{-9} f^{(1)} + 5.73438 \times 10^{-9} f^{(2)} - 1.6693 \times 10^{-9} f^{(3)}.$$

   Figure 8.11 illustrates the relative error

$$\frac{\|c^{(n)} - c\|}{\|c\|}$$

in the computation of the constant vector $c = (3, 2, -4)^\mathsf{T}$.

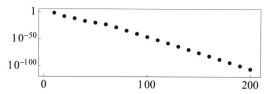

Fig. 8.11 The relative error $\|c^{(n)} - c\|/\|c\|$ for $N = 5, 10, \ldots, 200$

The plot shows that this relative error is much smaller than the corresponding relative error in the construction of $\psi^{(n)}$ as an approximation to $u|_{\partial S}$ (see Fig. 8.9), which is to be expected, given that Eq. (8.27) is very error-sensitive to the procedure regarding the computation of $c^{(n)}$.

The exponentially decreasing error is optimal in the sense that it is superior to that of any fixed-order method.

8.3 Numerical Example 2

With the same plate parameters, auxiliary circle ∂S_*, and points $x^{(k)}$ as in Sect. 8.2, and with

$$\sigma(x_p) = \begin{pmatrix} 10 & -4 & -1 + 3\cos(2\theta) \\ -4 & 10 & 1 - 3\cos(2\theta) \\ -1 + 3\cos(2\theta) & 1 - 3\cos(2\theta) & 13 + 3\cos(2\theta) \end{pmatrix},$$

we choose the boundary condition function $\mathscr{R}$ (in polar coordinates on ∂S) in (8.1) to be

$$\mathscr{R}(x_p)$$

$$= \frac{1}{2} \begin{pmatrix} -140 + 46\cos\theta + 3\sin\theta + 78\cos(2\theta) + 4\sin(2\theta) - 46\cos(3\theta) \\ \quad - 4\sin(3\theta) - 2\cos(4\theta) - 57\sin(4\theta) + 21\cos(5\theta) - 21\sin(5\theta) \\ 92 - 37\cos\theta - 34\sin\theta - 58\cos(2\theta) + 5\sin(2\theta) + 40\cos(3\theta) \\ \quad - 38\sin(3\theta) + 46\cos(4\theta) + 57\sin(4\theta) - 21\cos(5\theta) + 21\sin(5\theta) \\ 98 + 242\cos\theta - 36\sin\theta - 200\cos(2\theta) - 34\sin(2\theta) + 181\cos(3\theta) \\ \quad - 137\sin(3\theta) - 18\cos(4\theta) - \sin(4\theta) + 9\cos(5\theta) \\ \quad - 21\sin(5\theta) - 12\sin(6\theta) \end{pmatrix}.$$

It is easily verified that the exact solution u of (8.1) in this case is

$u(x)$

$$= \begin{pmatrix} 5 + (x_1^2 + x_2^2)^{-1}(-x_1) \\ \quad + (x_1^2 + x_2^2)^{-2}(-2x_1^2 - 4x_1 x_2 + 2x_2^2 + 2x_1^3 - 6x_1^2 x_2 \\ \quad + (x_1^2 + x_2^2)^{-3}(6x_1^2 x_2 - 2x_2^3 - 2x_1^4 + 8x_1^3 x_2 + 12x_1^2 x_2^2 - 8x_1 x_2^3 - 2x_2^4) \\ -1 + (x_1^2 + x_2^2)^{-1}(3x_2) \\ \quad + (x_1^2 + x_2^2)^{-2}(-8x_1 x_2 + 6x_1^2 x_2 - 2x_2^3 \\ \quad + (x_1^2 + x_2^2)^{-3}(-2x_1^3 + 6x_1 x_2^2 - 2x_1^4 - 8x_1^3 x_2 + 12x_1^2 x_2^2 + 8x_1 x_2^3 - 2x_2^4) \\ -4 + 5x_1 + x_2 - \tfrac{1}{2}\ln(x_1^2 + x_2^2) + (x_1^2 + x_2^2)^{-1}(-3x_1 - x_2 + 2x_1^2 - 2x_2^2) \\ \quad + (x_1^2 + x_2^2)^{-2}(6x_1^2 + 2x_1 x_2 - 6x_2^2 - x_1^3 + 3x_1^2 x_2 + 3x_1 x_2^2 - x_2^3) \\ \quad + (x_1^2 + x_2^2)^{-3}(-6x_1^3 + 18x_1^2 x_2 + 18x_1 x_2^2 - 6x_2^3) \end{pmatrix},$$

or, in polar coordinates,

$$u(x_p) = \begin{pmatrix} 5 + r^{-1}[-\cos\theta + 2\cos(3\theta)] \\ \quad + r^{-2}[2\cos(2\theta) - 2\sin(2\theta) - 2\cos(4\theta) + 2\sin(4\theta)] \\ \quad + r^{-3}[2\sin(3\theta)] \\ -1 + r^{-1}[3\sin\theta + 2\sin(3\theta)] \\ \quad + r^{-2}[-4\sin(2\theta) - 2\cos(4\theta) - 2\sin(4\theta)] \\ \quad + r^{-3}[-2\cos(3\theta)] \\ r(-5\cos\theta + \sin\theta) - 4 - 4\ln r + 2\cos(2\theta) \\ \quad + r^{-1}[-3\cos\theta - \sin\theta - \cos(3\theta) + \sin(3\theta)] \\ \quad + r^{-2}[6\cos(2\theta) + \sin(2\theta)] \\ \quad + r^{-3}[-6\cos(3\theta) + 6\sin(3\theta)] \end{pmatrix}. \quad (8.39)$$

8.9 Remark. Using this polar form and (1.15), we easily convince ourselves that solution (8.39) belongs to class $\mathscr{A}^*$, with the class $\mathscr{A}$ component $u^{\mathscr{A}}$ defined by the coefficients

$$m_0 = 0,$$
$$m_1 = -\tfrac{1}{2}, \quad m_2 = \tfrac{3}{2},$$
$$m_3 = -1, \quad m_4 = 0,$$
$$m_5 = -2, \quad m_6 = -1,$$
$$m_7 = 1, \quad m_8 = 0,$$
$$m_9 = 1, \quad m_{10} = 0,$$
$$m_{11} = 1, \quad m_{12} = 6.$$

The procedure described in Remark 1.11 also shows that this solution contains the rigid displacement component $f = 5f^{(1)} - f^{(2)} - 3f^{(3)}$, alternatively written as

$$f = Fc, \quad c = (5, -1, -3)^{\mathsf{T}}.$$

The details mentioned in Remarks 6.11–6.14 apply here as well.

We approximate the solution of problem (8.1) in this example by means of the MRR method.

8.3.1 Graphical Illustrations

The graphs of the components of $(u^{\mathscr{A}})^{(93)}$ computed from $\psi^{(93)}$ (that is, with 30 points $x^{(k)}$ on ∂S_*) for

$$r \geq 1.01, \quad 0 \leq \theta < 2\pi,$$

together with those of the components of $\psi^{(93)} - f$, which approximates $u^{\mathscr{A}}|_{\partial S}$, are shown in Fig. 8.12.

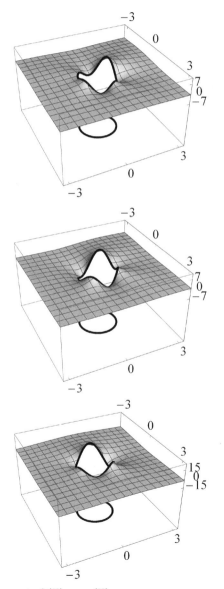

Fig. 8.12 Components of $(u^{\mathscr{A}})^{(93)}$ and $\psi^{(93)} - f$ (heavy lines around the edge of the hole)

The reason for the restriction $r \geq 1.01$ is the increasing influence of the singularities of $(D(x,y))^{\mathscr{A}}$ and $P(x,y)$ for $x \in S^-$ very close to $y \in \partial S$.

Figure 8.13 shows the graphs of the components of $(u^{\mathscr{A}})^{(93)}$ computed from $\psi^{(93)}$ for

$$1.01 \leq r \leq 100, \quad 0 \leq \theta < 2\pi,$$

illustrating the class $\mathscr{A}$ behavior of the solution away from the boundary. The vertical range of these plots has been truncated to allow better visualization.

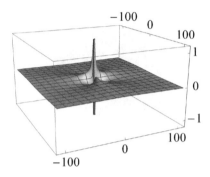

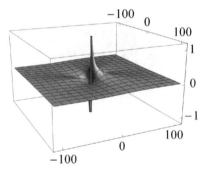

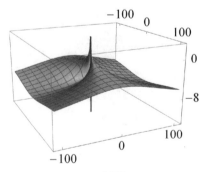

Fig. 8.13 Components of $(u^{\mathscr{A}})^{(93)}$ away from the boundary

The graphs of the components of the actual error $(u^{\mathscr{A}})^{(93)} - u^{\mathscr{A}}$ can be seen in Fig. 8.14. The approximation is between 2 and 4 digits of accuracy at a distance from ∂S; it deteriorates significantly close to the boundary, but improves considerably away from it.

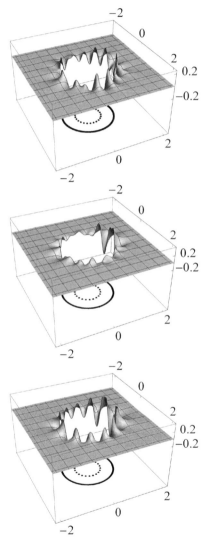

Fig. 8.14 Components of the error $(u^{\mathscr{A}})^{(93)} - u^{\mathscr{A}}$

8.3.2 Computational Procedure

The step-by-step outline of the computational procedure used for solving the exterior Robin problem starts with the boundary data function $\mathcal{R}$, whose components (in polar coordinates) are graphed in Fig. 8.15.

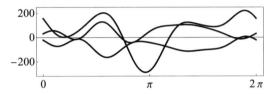

Fig. 8.15 Components of $\mathcal{R}$ in polar coordinates

$\mathcal{R}$ is digitized on ∂S and the functions $\varphi^{(i)}$ are constructed and digitized as described in Sect. 8.2.2. If instead of MRR we were to select an orthonormalization method—say, MGS—with the rigid displacement computed a priori as explained in Remark 8.6, then the $\omega^{(i)}$ would be digitized in the same way.

The graphs of typical three-component vector functions $\varphi^{(i)}$ and $\omega^{(i)}$ are displayed in Figs. 8.16 and 8.17, respectively. The increasing number of oscillations in the latter eventually produces computational difficulties in the evaluation of the inner products occurring in the construction of additional orthonormal vector functions.

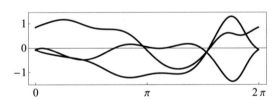

Fig. 8.16 Typical vector functions $\varphi^{(i)}$ in polar coordinates

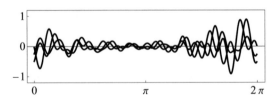

Fig. 8.17 Typical vector functions $\omega^{(i)}$ in polar coordinates

8.3.3 Computational Results

For the exterior Robin problem, the approximation $\psi^{(n)}$ of $u|_{\partial S}$ is computed by means of the system of equations generated by (8.26), (8.27), (8.30), and the $x^{(k)}$.

In our example, $\psi^{(93)}$ was calculated with 30 points $x^{(k)}$ on ∂S_* (see Fig. 8.18). As Fig. 8.19 shows, this approximation is quite good when compared with the exact result for $u|_{\partial S}$ derived from the known solution u.

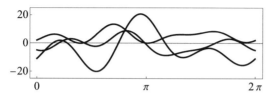

Fig. 8.18 Components of $\psi^{(93)}$ in polar coordinates

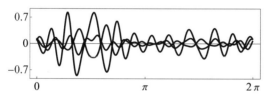

Fig. 8.19 Components of the error $\psi^{(93)} - u|_{\partial S}$ in polar coordinates

Equation (8.35) and $\psi^{(93)}$ produce the graphs displayed in Figs. 8.12 and 8.13.

8.3.4 Error Analysis

The approximation $\psi^{(n)}$, $n = 3N + 3$, is constructed with N equally spaced points $x^{(k)}$ on ∂S_*. The behavior of the relative error

$$\frac{\|\psi^{(3N+3)} - u|_{\partial S}\|}{\|u|_{\partial S}\|}$$

as a function of N determines the efficiency and accuracy of the computational procedure. The logarithmic plot in Fig. 8.20 illustrates the size of the relative error in terms of N.

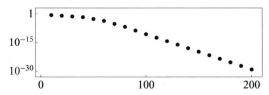

Fig. 8.20 The relative error $\|\psi^{(3N+3)} - u|_{\partial S}\|/\|u|_{\partial S}\|$ for $N = 5, 10, \ldots, 200$

This plot strongly suggests that the relative error improves exponentially as N increases. Fitting a linear curve to the logarithmic data for $N > 50$ yields the model

$$7.9817 - 0.18549 N.$$

A comparison of the logarithmic data and the linear model is shown in Fig. 8.21. The two are seen to be in good agreement for $N > 50$, so the relative error may be modeled by

$$\frac{\|\psi^{(3N+3)} - u|_{\partial S}\|}{\|u|_{\partial S}\|} = 9.5874 \times 10^7 \times 10^{-0.18549 N}$$

$$= 9.5874 \times 10^7 \times 0.652393^N.$$

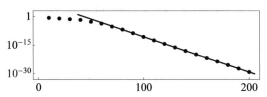

Fig. 8.21 The relative error $\|\psi^{(3N+3)} - u|_{\partial S}\|/\|u|_{\partial S}\|$ and the linear model for $N = 5, 10, \ldots, 200$

We remark that system (8.34) also generates the approximation

$$f^{(n)} = c_1^{(n)} f^{(1)} + c_2^{(n)} f^{(2)} + c_3^{(n)} f^{(3)},$$

or

$$f^{(n)} = F c^{(n)}, \quad c^{(n)} = (c_1^{(n)}, c_2^{(n)}, c_3^{(n)})^\mathsf{T},$$

of the rigid displacement

$$f = F c = 5 f^{(1)} - f^{(2)} - 3 f^{(3)}.$$

In our case, this is

$$f^{(93)} = 5.00074 f^{(1)} - 1.0031 f^{(2)} - 3.00029 f^{(3)},$$

with error

$$f^{(93)} - f = 0.00074 f^{(1)} - 0.0031 f^{(2)} - 0.00029 f^{(3)}.$$

Figure 8.22 illustrates the relative error

$$\frac{\|c^{(n)} - c\|}{\|c\|}$$

in the computation of the constant vector $c = (5, -1, -3)^{\mathsf{T}}$.

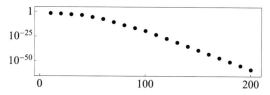

Fig. 8.22 The relative error $\|c^{(n)} - c\|/\|c\|$ for $N = 5, 10, \ldots, 200$

The plot shows that this relative error is much smaller than the corresponding relative error in the construction of $\psi^{(n)}$ as an approximation to $u|_{\partial S}$ (see Fig. 8.20), which is to be expected, given that Eq. (8.27) is very error-sensitive to the procedure regarding the computation of $c^{(n)}$.

The exponentially decreasing error is optimal in the sense that it is superior to that of any fixed-order method.

8.4 Numerical Example 3

With the same plate parameters, auxiliary circle ∂S_*, and points $x^{(k)}$ as in Sect. 8.2, and with

$$
\sigma(x) = \begin{pmatrix} 11 & 2 & 1 \\ 2 & 8 & -2 \\ 1 & -2 & 11 \end{pmatrix},
$$

we choose the boundary condition function $\mathscr{R}$ (in polar coordinates on ∂S) in (8.1) to be

$$
\mathscr{R}(x_p) = \begin{pmatrix} -27 + 52\cos\theta - 29\sin\theta + 12\cos(2\theta) - 27\sin(2\theta) \\ \quad - 37\cos(3\theta) + 23\sin(3\theta) - 4\cos(4\theta) + 21\sin(4\theta) \\ 30 - 14\cos\theta + 4\sin\theta + 43\cos(2\theta) - 10\sin(2\theta) \\ \quad - 22\cos(3\theta) - 34\sin(3\theta) - 15\cos(4\theta) + 4\sin(4\theta) \\ -33 + 64\cos\theta - 77\sin\theta - 116\cos(2\theta) + 68\sin(2\theta) \\ \quad + \cos(3\theta) + 67\sin(3\theta) + 4\cos(4\theta) + 2\sin(4\theta) \end{pmatrix}.
$$

The exact solution u of the problem is not known in this case.

The details mentioned in Remarks 6.11–6.14 apply here as well.

We approximate the solution of (8.1) in this example by means of the MRR method.

8.4.1 Graphical Illustrations

The graphs of the components of $(u^{\mathscr{A}})^{(63)}$ computed from $\psi^{(63)}$ (that is, with 20 points $x^{(k)}$ on ∂S_*) by means of (8.36) for

$$
r \geq 1.01, \quad 0 \leq \theta < 2\pi,
$$

together with those of the components of $\psi^{(63)} - f^{(63)}$, which approximates $u^{\mathscr{A}}|_{\partial S}$, are shown in Fig. 8.23.

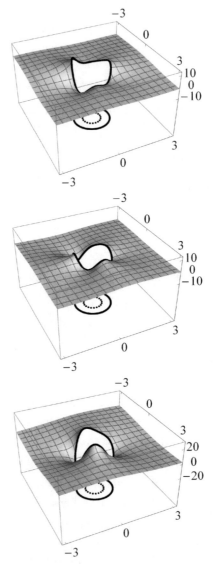

Fig. 8.23 Components of $(u^{\mathscr{A}})^{(63)}$ and $\psi^{(63)} - f^{(63)}$ (heavy lines around the edge of the hole)

Figure 8.24 contains the graphs of the components of $(u^{\mathscr{A}})^{(63)}$ computed from $\psi^{(63)}$ for

$$1.01 \le r \le 100, \quad 0 \le \theta < 2\pi,$$

illustrating the class $\mathscr{A}$ behavior of the solution away from the boundary. The vertical range of these plots has been truncated to allow better visualization.

The reason for the restriction $r \ge 1.01$ is the increasing influence of the singularities of $(D(x,y))^{\mathscr{A}}$ and $P(x,y)$ for $x \in S^-$ very close to $y \in \partial S$.

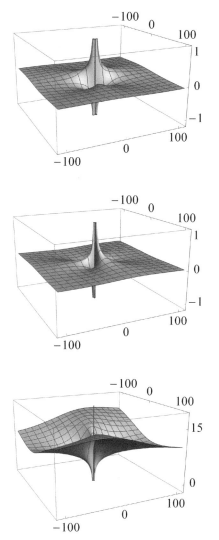

Fig. 8.24 Components of $(u^{\mathscr{A}})^{(63)}$ away from the boundary

8.4.2 Computational Procedure

The step-by-step outline of the computational procedure used for solving the exterior Robin problem starts with the boundary data function $\mathscr{R}$, whose components (in polar coordinates) are graphed in Fig. 8.25.

$\mathscr{R}$ is now digitized on ∂S and the functions $\varphi^{(i)}$ are constructed and digitized as described in Sect. 8.2.2.

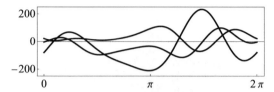

Fig. 8.25 Components of $\mathscr{R}$ in polar coordinates

8.4.3 Computational Results

For the exterior Robin problem, the approximation $\psi^{(n)}$ of $u|_{\partial S}$ is computed by means of the system of equations generated by (8.26), (8.27), (8.30), and the $x^{(k)}$.

In our example, $\psi^{(63)}$ (see Fig. 8.26) was calculated with 20 points $x^{(k)}$ on ∂S_*. This approximation was then combined with $(u^{\mathscr{A}})^{(63)}$ to generate the graphs in Fig. 8.23, which show good agreement between $(u^{\mathscr{A}})^{(63)}$ and $\psi^{(63)} - f^{(63)}$. This is an indirect validation of our method.

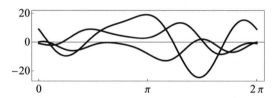

Fig. 8.26 Components of $\psi^{(63)}$ in polar coordinates

Apart from forming the basis for the approximation $\psi^{(n)}$, system (8.34) also generates the approximation

$$f^{(n)} = c_1^{(n)} f^{(1)} + c_2^{(n)} f^{(2)} + c_3^{(n)} f^{(3)},$$

or

$$f^{(n)} = Fc^{(n)}, \quad c^{(n)} = (c_1^{(n)}, c_2^{(n)}, c_3^{(n)})^{\mathsf{T}},$$

of the rigid displacement $f = Fc$. In our case, this is

$$f^{(63)} = 3.000000003295\, f^{(1)} - 4.000000008200\, f^{(2)} + 1.54246 \times 10^{-9} f^{(3)}.$$

When the rigid displacement f is computed with 200 points $x^{(k)}$, the error for $f^{(63)}$ can be approximated by

$$f^{(63)} - f^{(603)} = 3.2956 \times 10^{-9} f^{(1)} - 8.2004 \times 10^{-9} f^{(2)} + 0.5426 \times 10^{-9} f^{(3)}.$$

8.4.4 Error Analysis

For the exterior Robin problem considered in this example, the approximation $\psi^{(n)}$, $n = 3N + 3$, of $u|_{\partial S}$ is constructed with N equally spaced points $x^{(k)}$ on ∂S_*. The behavior of the relative error

$$\frac{\|\psi^{(3N+3)} - u|_{\partial S}\|}{\|u|_{\partial S}\|}$$

as a function of N determines the efficiency and accuracy of the computational procedure. However, since the exact expression of $u|_{\partial S}$ is not known in this example, an accurate error analysis cannot be performed.

As a substitute, we propose instead an indirect route to obtaining a measure of partial validation of our calculations. We use $\psi^{(63)} = (u|_{\partial S})^{(63)}$ as the boundary condition for an exterior Dirichlet problem and use the approximate solution of the latter to compute (by means of the procedure laid out in Chap. 6) an approximation $\Gamma^{(63)}$ of Tu (on ∂S), which, if our method is sound, should lead to a close fit to $\mathscr{R}$. Figure 8.27 shows the components of $\Gamma^{(63)}$.

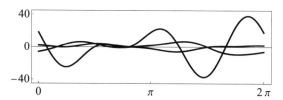

Fig. 8.27 Components of $\Gamma^{(63)}$ in polar coordinates

Combining $\psi^{(63)}$ and $\Gamma^{(63)}$ yields the approximation of the Robin boundary condition $\mathscr{R}^{(63)} = \Gamma^{(63)} + \sigma \psi^{(63)}$.

The components of $\mathscr{R}^{(63)}$ and those of the difference $\mathscr{R}^{(63)} - \mathscr{R}$ are shown in Figs. 8.28 and 8.29, respectively.

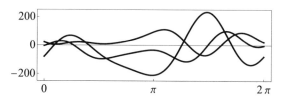

Fig. 8.28 Components of $\mathscr{R}^{(63)}$ in polar coordinates

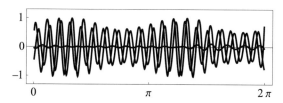

Fig. 8.29 Components of $\mathscr{R}^{(63)} - \mathscr{R}$ in polar coordinates

Figure 8.29 indicates that the total combined error in this verification exercise is consistent with the error expected when we use MRR to compute $\mathscr{R}^{(63)}$. Repeating our experiment with 200 points $x^{(k)}$ instead of 20 on ∂S_*, we obtained the much smaller relative error

$$\frac{\|\mathscr{R}^{(603)} - \mathscr{R}\|}{\|\mathscr{R}\|} = 5.69512 \times 10^{-52}.$$

This confirms the robust nature of the generalized Fourier series method of approximation presented here.

8.5 Numerical Example 4

With the same plate parameters, auxiliary circle ∂S_*, and points $x^{(k)}$ as in Sect. 8.2, and with

$$\sigma(x_p) = \begin{pmatrix} 10 & -4 & -1 + 3\cos(2\theta) \\ -4 & 10 & 1 - 3\cos(2\theta) \\ -1 + 3\cos(2\theta) & 1 - 3\cos(2\theta) & 13 + 3\cos(2\theta) \end{pmatrix},$$

we choose the boundary condition function $\mathscr{R}$ (in polar coordinates on ∂S) in (8.1) to be

$\mathscr{R}(x_p)$

$$= \frac{1}{2} \begin{pmatrix} -27 - 67\cos\theta - 26\sin\theta + 44\cos(2\theta) + 18\sin(2\theta) \\ \quad + 93\cos(3\theta) - 16\sin(3\theta) - 149\cos(4\theta) + 85\sin(4\theta) - 84\cos(5\theta) \\ 39 + 47\cos\theta + 6\sin\theta - 12\cos(2\theta) - 50\sin(2\theta) \\ \quad - 81\cos(3\theta) + 40\sin(3\theta) + 61\cos(4\theta) - 173\sin(4\theta) + 84\cos(5\theta) \\ -57 - 47\cos\theta - 42\sin\theta - 42\cos(2\theta) + 174\sin(2\theta) \\ \quad - 579\cos(3\theta) - 8\sin(3\theta) + 31\cos(4\theta) + 17\sin(4\theta) - 78\cos(5\theta) \\ \quad - 6\sin(5\theta) - 24\cos(6\theta) + 24\sin(6\theta) \end{pmatrix}.$$

The exact solution u of the problem is not known in this case.

The details mentioned in Remarks 6.11–6.14 apply here as well.

We approximate the solution of (8.1) in this example by means of the MRR method.

8.5.1 Graphical Illustrations

The graphs of the components of $(u^{\mathscr{A}})^{(93)}$ computed from $\psi^{(93)}$ (that is, with 30 points $x^{(k)}$ on ∂S_*) by means of (8.36) for

$$r \geq 1.01, \quad 0 \leq \theta < 2\pi,$$

together with those of the components of $\psi^{(93)} - f^{(93)}$, which approximates $u^{\mathscr{A}}|_{\partial S}$, are shown in Fig. 8.30.

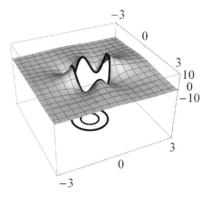

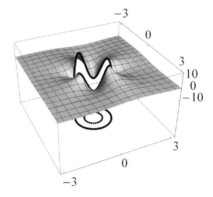

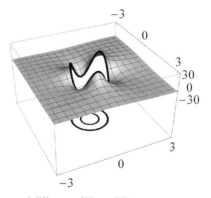

Fig. 8.30 Components of $(u^{\mathscr{A}})^{(93)}$ and $\psi^{(93)} - f^{(93)}$ (heavy lines around the edge of the hole)

The reason for the restriction $r \geq 1.01$ is the increasing influence of the singularities of $(D(x,y))^{\mathscr{A}}$ and $P(x,y)$ for $x \in S^-$ very close to $y \in \partial S$.

Figure 8.31 contains the graphs of the components of $(u^{\mathscr{A}})^{(93)}$ computed from $\psi^{(93)}$ for

$$1.01 \leq r \leq 100, \quad 0 \leq \theta < 2\pi,$$

illustrating the class $\mathscr{A}$ behavior of the solution away from the boundary. The vertical range of these plots has been truncated to allow better visualization.

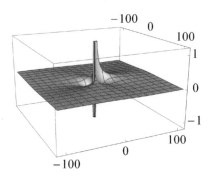

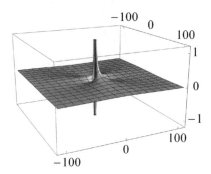

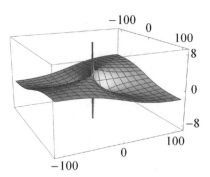

Fig. 8.31 Components of $(u^{\mathscr{A}})^{(93)}$ away from the boundary

8.5.2 Computational Procedure

The step-by-step outline of the computational procedure used for solving the exterior Robin problem starts with the boundary data function $\mathscr{R}$, whose components (in polar coordinates) are graphed in Fig. 8.32.

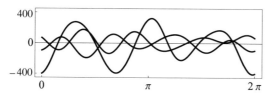

Fig. 8.32 Components of $\mathscr{R}$ in polar coordinates

$\mathscr{R}$ is now digitized on ∂S and the functions $\varphi^{(i)}$ are constructed and digitized as described in Sect. 8.2.2.

8.5.3 Computational Results

For the exterior Robin problem, the approximation $\psi^{(n)}$ of $u|_{\partial S}$ is computed by means of the system of equations generated by (8.26), (8.27), (8.30), and the $x^{(k)}$.

In our example, $\psi^{(93)}$ (see Fig. 8.33) was calculated with 30 points $x^{(k)}$ on ∂S_*. This approximation was then combined with $(u^{\mathscr{A}})^{(93)}$ to generate the graphs in Fig. 8.30, which show good agreement between $(u^{\mathscr{A}})^{(93)}$ and $\psi^{(93)} - f^{(93)}$. This is an indirect validation of our method.

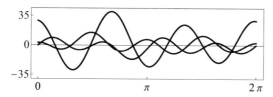

Fig. 8.33 Components of $\psi^{(93)}$ in polar coordinates

Apart from forming the basis for the approximation of $\psi^{(n)}$, system (8.34) also generates the approximation

$$f^{(n)} = c_1^{(n)} f^{(1)} + c_2^{(n)} f^{(2)} + c_3^{(n)} f^{(3)},$$

or

$$f^{(n)} = F c^{(n)}, \quad c^{(n)} = (c_1^{(n)}, c_2^{(n)}, c_3^{(n)})^{\mathsf{T}},$$

of the rigid displacement $f = Fc$. In our case, this is

$$f^{(93)} = 0.997177 f^{(1)} - 1.99812 f^{(2)} + 3.00054 f^{(3)}.$$

When the rigid displacement f is computed with 200 points $x^{(k)}$, the error for $f^{(93)}$ can be approximated by

$$f^{(93)} - f^{(603)} = -0.0028231 f^{(1)} + 0.0018812 f^{(2)} + 0.00054103 f^{(3)}.$$

8.5.4 Error Analysis

The approximation $(u|_{\partial S})^{(n)} = \psi^{(n)}$, $n = 3N + 3$, is constructed with N equally spaced points $x^{(k)}$ on ∂S_*. The behavior of the relative error

$$\frac{\|\psi^{(3N+3)} - u|_{\partial S}\|}{\|u|_{\partial S}\|}$$

as a function of N determines the efficiency and accuracy of the computational procedure. However, since the exact expression of $u|_{\partial S}$ is not known in this example, an accurate error analysis cannot be performed.

As a substitute, we propose instead an indirect route to obtaining a measure of partial validation of our calculations. We use $\psi^{(93)} = (u|_{\partial S})^{(93)}$ as the boundary condition for an exterior Dirichlet problem and use the approximate solution of the latter to compute (by means of the procedure laid out in Chap. 6) an approximation $\Gamma^{(93)}$ of Tu (on ∂S), which, if our method is sound, should lead to a close fit to $\mathscr{R}$. Figure 8.34 shows the components of $\Gamma^{(93)}$.

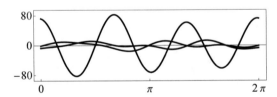

Fig. 8.34 Components of $\Gamma^{(93)}$ in polar coordinates

Combining $\psi^{(93)}$ and $\Gamma^{(93)}$ yields the approximation of the Robin boundary condition function

$$\mathscr{R}^{(93)} = \Gamma^{(93)} + \sigma \psi^{(93)}.$$

The components of $\mathscr{R}^{(93)}$ and those of the difference $\mathscr{R}^{(93)} - \mathscr{R}$ are shown in Figs. 8.35 and 8.36, respectively.

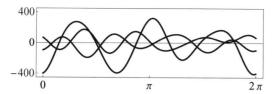

Fig. 8.35 Components of $\mathscr{R}^{(93)}$ in polar coordinates

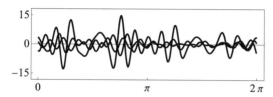

Fig. 8.36 Components of $\mathscr{R}^{(93)} - \mathscr{R}$ in polar coordinates

Figure 8.36 indicates that the total combined error in this verification exercise is consistent with the error expected when we use MRR to compute $\mathscr{R}^{(93)}$. Repeating our experiment with 200 points $x^{(k)}$ instead of 30 on ∂S_*, we obtained the much smaller relative error

$$\frac{\|\mathscr{R}^{(603)} - \mathscr{R}\|}{\|\mathscr{R}\|} = 6.78631 \times 10^{-28}.$$

This confirms the robust nature of the generalized Fourier series method of approximation presented here.

Appendix A
Numerical Issues

In the preceding chapters, we supplied the numerical computation and analysis for each of the six boundary value problems treated by the generalized Fourier series method. Part of that analysis was determining how much error was contained in the final answer. It is convenient to subdivide the total error into two contributing components: algorithmic computational error and floating-point representational error. The former was covered in detail in earlier chapters. Our aim here is to give an abbreviated summary of the issues involved in understanding these two components, by means of the *Mathematica*® software. For this purpose, we intend to express our analysis and results in the *Mathematica*® language notation, which allows both *Mathematica*® code and computations to be displayed in a uniform format. For example, a function $f(x)$ is expressed in *Mathematica*® notation as $f[x]$, and a vector $(x_1, x_2, x_3, \ldots, x_n)$ is written as $\{x_1, x_2, x_3, \ldots, x_n\}$.

A.1 Arbitrary Precision Floating-Point Numbers

Mathematica® allows the use of exact numbers—for example, integers and rational numbers—which are considered to have infinite precision; thus,

$$\text{Precision}[1/3] \to \infty.$$

Approximate numbers can be constructed to an arbitrarily specified level of precision. For example, the exact number 1/3 expressed as an approximate floating-point number to 20 digits of precision is

$$N[1/3, 20] \to 0.33333333333333333333.$$

A complete analysis of arbitrary precision floating-point computational error involves controlling both the number of digits specified by the *Accuracy* and the num-

© Springer Nature Switzerland AG 2020
C. Constanda, D. Doty, *The Generalized Fourier Series Method*, Developments in
Mathematics 65, https://doi.org/10.1007/978-3-030-55849-9

ber of digits specified by the *Precision* parameters. Controlling the magnitude of the total error in a computed numerical result x requires the total error to be less than

$$10^{-\text{Accuracy}} + |x|10^{-\text{Precision}}.$$

This inequality is true for our specific example irrespective of the *Accuracy*; that is,

$$1/3 - 0.33333333333333333333 \leq 10^{-\text{Accuracy}} + |1/3|10^{-20}.$$

A.2 Computational Error by Means of Arbitrary Precision Floating-Point Numbers

We restrict this brief explanation to examining the influence of precision. Simply stated, the precision represents the number of significant digits in the mantissa of a floating-point number. The reader is referred to the extensive documentation in *Mathematica*® for a more complete description.

A detailed analysis of computational error is extremely involved and intricate. As such, it is well beyond the scope of this book, and is to be replaced by a few simple examples that illustrate the fundamental underlying ideas.

Consider the problem of numerically evaluating a one-variable function $f[x]$ using floating-point computations. The function f is known, and the exact value of x has been replaced by a floating-point approximation x^* (typically the result of a previous numerical calculation), which has been computed to a known precision. Our objective is to compute $f[x]$, if possible, to the same degree of precision when the exact value of x is unknown but is close to x^*.

The exact value of $f[x]$ can be approximated by

$$f[x] \simeq f[x^*] + f'[x^*](x - x^*).$$

If x^* has a known or chosen precision of Precision_1 digits by reference to the exact number x, the magnitude of the error in x is

$$x - x^* \simeq |x^*|10^{-\text{Precision}_1}.$$

If $f'[x^*]$ is reasonably well behaved, then the magnitude of the error in $f[x]$ is

$$f[x] - f[x^*] \simeq |f'[x^*]| \, |x^*|10^{-\text{Precision}_1},$$

which has the same precision magnitude as x^*. To prevent any further loss of precision when computing the approximation $(f[x^*])^*$ to $f[x^*]$, it is clear that we should compute $f[x^*]$ to a precision of Precision_2 digits that is the same or better than Precision_1 digits:

$$f[x^*] - (f[x^*])^* \simeq |f[x^*]|10^{-\text{Precision}_2}.$$

Therefore, the precision magnitude in computing the approximation $(f[x^*])^*$ to $f[x]$ is

$$f[x] - (f[x^*])^* \simeq |f'[x^*]||x^*|10^{-\text{Precision}_1} + |f[x^*]|10^{-\text{Precision}_2}.$$

This equals the minimum of Precision_1 and Precision_2, which becomes Precision_1 digits, thus preserving the final magnitude of precision.

For example, if

$$x = 1/3, \text{ usually not specifically known,}$$
$$\text{Precision}_1 = 5 \text{ digits, which is known,}$$
$$x^* = 0.33333, \text{ also known,}$$
$$f = \text{Tan,}$$

then the exact value of x is approximated by x^* with a precision of 5 digits. The exact value of $\text{Tan}[1/3]$ is approximated by

$$\text{Tan}[1/3] - \text{Tan}[0.33333] \simeq \text{Tan}'[0.33333](1/3 - 0.33333),$$

which also has a precision of 5 digits. The exact value of $\text{Tan}[0.33333]$ is approximated by the computed value $(\text{Tan}[0.33333])^*$, where

$$(\text{Tan}[0.33333])^* = 0.34624981654314882570986281249957090586.$$

We notice that $(\text{Tan}[0.33333])^*$ was computed internally with a precision of $\text{Precision}_2 = 38$ digits, 33 digits more than Precision_1. This is typical of a functional evaluation, where the internal computational precision is sufficiently augmented to guarantee a computed result that meets or exceeds the final desired precision. This procedure is required by the inevitable deterioration of computational accuracy. The exact amount of additional internal computational precision is determined by $Mathematica^{®}$ and depends on the computed function, the precision of x^*, and the floating-point processor used. Hence, the exact value of $\text{Tan}[1/3]$ is

$$\text{Tan}[1/3] = 0.34624981654314882570986281249957090586 + \text{error}_1 + \text{error}_2,$$

where error_1 is about 38 digits smaller than $\text{Tan}[1/3]$, and error_2 is about 5 digits smaller than $\text{Tan}[1/3]$. After rounding and truncating to indicate significant digits, we have

$$\text{Tan}[1/3] = 0.34625 + \text{error}_3,$$

where error_3 is also about 5 digits smaller than $\text{Tan}[1/3]$, thus preventing any significant deterioration in the final computed precision of 5 digits.

This procedure becomes especially important when the function $f[x]$ is very ill-conditioned. For example, if x^* has a precision of 10 digits and we know that the floating-point computation of $(f[x^*])^*$ produces a loss of 20 digits of precision owing to ill-conditioning, then the floating-point computation of $(f[x^*])^*$ needs to have a target exceeding 30 digits of precision in order to maintain the input precision of 10 digits.

In reality, *Mathematica*® does not round and truncate intermediate computed results. They are used in their totality in subsequent computations. However, the current level of precision is appended to the result so that the degree of accuracy can be monitored throughout a lengthy computation. For example, the actual output in our case would contain both the computed number and an indication of the 5.0 significant digits of precision that it represents, appended at the end:

$$(\text{Tan}[0.33333])^* = 0.3462498165431488257098628124995709058656\text{`}5.0.$$

This observation is important in our considerations. *Mathematica*® is designed to ensure that the reported level of precision is absolutely guaranteed, and to do so, it tends to be 'pessimistic' in the sense that the actual computation typically results in a higher than reported precision. When confronted with a massive amount of computation, the reported degree of precision can be dramatically under-reported, as happens in our case, which therefore requires special attention on our part.

A.3 Machine-Precision Numbers

Floating-point numbers can be constructed using an arbitrary value for the precision. This is usually done when the default machine precision is insufficient to perform the required calculation. However, machine-precision calculations are typically faster because much of the computation is performed within the hardware rather than the software. There are consequences and limitations imposed by the use of machine-precision numbers. The first is that the precision is limited to that built into the floating-point processor. A typical value corresponds to a 64 binary bit floating-point number, which has a mantissa with precision of approximately 16 digits:

$$\$\text{MachinePrecision} \rightarrow 15.954589770191003.$$

A second limitation is that no attempt is made to monitor the deterioration of precision during a sequence of computations. As a consequence, there is no direct indication of the precision contained within the final answer. The final answer, as well as each intermediate step, are always assumed to have the fixed machine precision, whether or not that level of precision is justified.

A third limitation is that machine-precision computation is dependent upon the specific floating-point processor being used, along with its supporting microcode contained within the hardware. As a consequence, different machines using the same upper-level *Mathematica*® language can produce different results. There is no way of using machine-precision calculations in a totally unambiguous manner.

We give a brief outline of the evaluation of a simple one-variable function $f[x]$ using machine-precision computation. The function $f[x]$ represents one step in a sequence of steps needed to compute the final answer. Also, $f[x]$ is to be evaluated internally within the *Mathematica*® language as a single unit, and is not fabricated by the user from smaller pieces. We start by truncating the exact number x into a machine-precision number x^*. *Mathematica*® would also like to compute $(f[x^*])^*$ from $f[x^*]$ to machine precision. The reason for this is to ensure as little deterioration in the final answer as possible. Typically, this cannot be achieved by exclusively using only machine precision for the internal computation. Consequently, *Mathematica*® temporarily augments the computational precision beyond machine precision during this calculation. The exact amount of additional precision used is determined within the software and depends on the function being evaluated. However, for a target precision of about 16 digits, at most 50 digits of additional precision is usually more than sufficient. Once $(f[x^*])^*$ is computed using this additional precision, the final number is rounded and truncated to machine-precision digits. This intermediate result is then transferred to the next step.

A.4 Fixed-Precision Floating-Point Numbers

When trying to separate the effect of floating-point representational error from algorithmic computational error, it is necessary for us to reduce the size of one of them so that the error being analyzed becomes significantly dominant, which makes it easy to identify and measure. For this reason, when trying to determine the magnitude of the algorithmic computational error in earlier chapters, we found it desirable to make the contribution caused by floating-point numbers significantly smaller. Such a result can be achieved by increasing the floating-point representational precision from machine precision to a significantly higher number of digits, for example, 50 or 100 digits.

This is done by switching to arbitrary precision floating-point numbers. But, as described above, the internal procedure used in arbitrary precision evaluations is quite different from that required when using fixed machine precision. Specifically, arbitrary precision calculations monitor the steadily deteriorating level of precision throughout the process, decreasing the required precision for each subsequent step. This makes it desirable to simulate a fixed-precision procedure using arbitrary precision floating-point numbers. The procedure is straightforward and follows that outlined for machine-precision floating-point calculations.

We start by choosing a fixed value of precision for our computation, for example, 50 or 100 digits. Next, we consider the evaluation of a simple one-variable function $f[x]$. This function represents one step in a sequence of steps needed to compute the final answer. Also, $f[x]$ is to be evaluated internally within *Mathematica*® as a single unit, and is not put together by the user from smaller pieces. We begin by explicitly adding additional digits or truncating the number x to a fixed-precision number x^*, constructed to have, and assigned to have, the chosen precision. We remark that the assigned precision is typically higher than the actual precision, as happens in the procedure used in machine-precision floating-point numbers. We then compute $(f[x^*])^*$ from $f[x^*]$ using *Mathematica*® . Since x^* is specified as having the desired fixed precision, *Mathematica*® attempts to compute $(f[x^*])^*$ from $f[x^*]$ to the same or higher degree of precision. To this end, *Mathematica*® temporarily augments the computational precision beyond the chosen precision during calculation. Once $(f[x^*])^*$ is computed with this additional precision, the final number is augmented, or is rounded and truncated, to have the chosen fixed precision. This intermediate result is then transferred to the next step, and the process is repeated.

One consequence of the above procedure is that it unavoidably slows the computational process. However, it can do a good job of isolating the effect of algorithmic computational error from floating-point representational error, which is our objective. For example, suppose that the magnitude of the algorithmic computational error is the one desired for analysis and measurement. Also, suppose that numerical experimentation indicates that the algorithmic computational error will contaminate the final answer by decreasing its precision from exact to only 30 digits. Additionally, suppose that we expect the fixed-precision floating-point computation to contaminate the final answer by decreasing the precision from the chosen input floating-point precision by at most 10 digits. Then, choosing a fixed floating-point precision of 50 digits, we have sufficient guarantee that the floating-point representational error does not contaminate the first $50 - 10 = 40$ digits in the computed answer. Since the algorithmic computational error leaves only the first 30 digits uncontaminated, this effectively isolates the floating-point representational error from the dominant algorithmic computational error.

Determining the loss of precision that occurs during a fixed computation because of floating-point representational error is also relatively straightforward. For example, to establish the loss of precision for a specific computation using 50 digit floating-point numbers, we perform the same calculation twice, once with 50 digits and again with 75 digits of precision. The loss of precision from 50 digits can now be found by examining where the 50-digit computed result begins to differ from the more accurate 75-digit outcome.

Appendix B
Numerical Integration

B.1 Introduction

The rest of the Appendix computes and analyzes examples motivated by the Generalized Fourier Series method developed earlier. The computation requires the ability to both algebraically manipulate functions and evaluate the integral representation formula, functional inner products, and norms. It will therefore be necessary for us to choose a method for numerically approximating the functions occurring in all of these computations. Whatever method we select, it is highly desirable to have it accurately compute the definite integrals generated by the integral equations concerned. Many methods are available for this—for example, power series approximations, Fourier series approximations, polynomial spaces, piecewise polynomial splines, and so on. Each of these methods has its own advantages and disadvantages. However, they are all constrained to have a fixed order of approximation once they are constructed. The natural choice of a quadrature rule for this specific application is the trapezoidal rule. It is well known that for the special case of an infinitely smooth periodic function, the trapezoidal rule has an exceptionally high order of convergence. Consequently, it would be unwise to pick a method for approximating our functions that imposes a limit on the overall order of accuracy. This automatically eliminates all the other methods mentioned above.

Consider one of the boundary value problems developed earlier. This problem is defined in a domain with a closed boundary curve. The accuracy of the numerical quadrature rule used to evaluate the integrals along the boundary is limited by the smoothness of the integrand and that of the boundary curve. The method developed here requires both the parametrized function $f[t]$ and boundary curve to be infinitely smooth. Hence, we consider the general problem of numerically computing a definite integral of the form

$$\int_a^b f[t]\,dt,$$

© Springer Nature Switzerland AG 2020
C. Constanda, D. Doty, *The Generalized Fourier Series Method*, Developments in
Mathematics 65, https://doi.org/10.1007/978-3-030-55849-9

where the function $f[t]$ is infinitely smooth and periodic with period $b - a$. The quadrature rule chosen for evaluation is the trapezoidal rule, which is known to exhibit optimal order of convergence for this special case. The trapezoidal rule that we choose to implement uses equally spaced quadrature nodes. Thus, $f[t]$ is approximated by a discretized function interpolating f at these nodes.

B.2 Discretized Functions

Let $f[t]$, $a \leq t \leq b$, be a periodic function, let n be a positive integer, and let $h = (b - a)/n$ and $t_j = a + jh$, $0 \leq j \leq n$. Then the discretized function $\Delta_n[f]$ is the n component vector that interpolates $f[t]$ at each t_j; that is,

$$\Delta_n(f) = \{f[t_0], f[t_1], f[t_2], f[t_3], \ldots, f[t_{n-1}]\}.$$

We recall that $f[t_n] = f[t_0]$ because f is periodic.

The algebraic operations for two functions f and g can now easily be computed from their discretized representation. For example,

$$\Delta_n(f + g) = \Delta_n(f) + \Delta_n(g),$$
$$\Delta_n(f * g) = \Delta_n(f) * \Delta_n(g),$$
$$\Delta_n(f \operatorname{div} g) = \Delta_n(f) / \Delta_n(g),$$

where the vector operations are interpreted as list operations in *Mathematica*® . If the vector function $\Delta_n(f)[t]$ is to be evaluated at t, where t is between the interpolation nodes, then an appropriate approximation (for example, a piecewise polynomial spline) can be constructed from $\Delta_n(f)$ for this purpose. Also, if a definite integral is to be evaluated, then we use the trapezoidal rule with the nodes t_j.

B.3 Trapezoidal Quadrature Rule

Again, we consider a periodic function $f[t]$, $a \leq t \leq b$, a positive integer n, and $h = (b - a)/n$, $t_j = a + jh$, $0 \leq j \leq n$. Then the trapezoidal rule on equally spaced nodes is defined by

$$T_n(f) = h\left(\tfrac{1}{2} f[t_0] + f[t_1] + f[t_2] + f[t_3] + \cdots + f[t_{n-1}] + \tfrac{1}{2} f[t_n]\right),$$

which converted into periodic discretized notation becomes

$$T_n(\Delta_n(f)) = h \sum_{j=0}^{n-1} \Delta_n(f)[t_j].$$

Hence, we have the discrete approximations

$$\Delta_n \left(\int_a^b f[t] \, dt \right) = T_n(\Delta_n(f)),$$

$$\Delta_n(< f, g >) = T_n(\Delta_n(f) * \Delta_n(g)),$$

$$\Delta_n(\|f\|_2) = \sqrt{T_n\left((\Delta_n(f))^2\right)},$$

$$\Delta_n(\|f\|_\infty) = \mathrm{Max}|\Delta_n(f)|,$$

where $\|\cdot\|_2$ and $\|\cdot\|_\infty$ are the L^2-norm and the L^∞-norm, respectively. These expressions are extremely easy to evaluate in *Mathematica*®.

B.4 Euler–Maclaurin Summation Formula

It is convenient to determine the accuracy of the trapezoidal rule for our special situation. There are many estimates corresponding to the optimal error for specific cases in the literature. However, for our purpose it suffices to derive the error from the Euler–Maclaurin summation formula.

The Bernoulli numbers b_k, needed in the construction of the Euler–Maclaurin summation formula, are calculated from the Bernoulli polynomials $B_k[t]$ at $t = 0$, where $b_k = B_k[0]$. The $B_k[t]$ on the interval $0 \leq t \leq 1$ satisfy, for all $k \geq 1$, the recursive differential equation

$$B'_k[t] = kB_{k-1}[t],$$

where

$$B_0[t] = 1,$$

$$\int_0^1 B_k[t] \, dt = 0,$$

$$b_k = B_k[0];$$

therefore,

$$b_0 = 1, \quad b_1 = -1/2, \quad b_2 = 1/6, \quad b_4 = -1/30, \quad b_6 = 1/42, \quad \ldots$$

and

$$b_{2k+1} = 0.$$

If m and n are positive integers, $h = (b-a)/n$, and $f[t]$ is $(2m+2)$-times continuously differentiable, then the Euler–Maclaurin summation formula is

$$\int_a^b f[t]\,dt = T_n\big(\Delta_n(f)\big) - F_{m,n} - E_{m,n},$$

where h is a function of n and

$$F_{m,n} = \sum_{i=1}^m \frac{h^{2i}b_{2i}}{(2i)!}\big(f^{(2i-1)}[b] - f^{(2i-1)}[a]\big),$$

$$E_{m,n}[\xi] = \frac{h^{2m+2}(b-a)b_{2m+2}f^{(2m+2)}[\xi]}{(2m+2)!},$$

b_{2m+2} are the Bernoulli numbers, and there are real numbers $\xi^*_{m,n}$, $a \le \xi^*_{m,n} \le b$, such that

$$E_{m,n} = E_{m,n}[\xi^*_{m,n}].$$

B.5 Trapezoidal Quadrature Rule: Error Analysis

The total error when using the trapezoidal rule for positive integers m and n is

$$T_n\big(\Delta_n(f)\big) - \int_a^b f[t]\,dt = F_{m,n} + E_{m,n}.$$

Consider the special case where $f[t]$ is periodic on the interval $a \le t \le b$ and all its derivatives of all orders are continuous. For each positive integer m, we have $F_{m,n} = 0$, and the corresponding error for the trapezoidal rule reduces to

$$T_n\big(\Delta_n(f)\big) - \int_a^b f[t]\,dt = E_{m,n} = \frac{h^{2m+2}(b-a)b_{2m+2}f^{(2m+2)}[\xi^*_{m,n}]}{(2m+2)!}.$$

This equality must hold for all positive values of m for each fixed n. Consequently, for each positive integer n, the error estimate becomes

$$\left| T_n\big(\Delta_n(f)\big) - \int_a^b f[t]\,dt \right| \le \underset{m\ge1}{\mathrm{Min}}|E_{m,n}| \le \underset{m\ge1}{\mathrm{Min}}\|E_{m,n}[\xi]\|_\infty$$

$$= \underset{m\ge1}{\mathrm{Min}}\frac{h^{2m+2}(b-a)|b_{2m+2}|\|f^{(2m+2)}\|_\infty}{(2m+2)!}.$$

In our problem, $\|f^{(2m+2)}\|_\infty$ tends to grow quite quickly as m increases. This is countered by the fact that $h^{2m+2}(b-a)|b_{2m+2}|/(2m+2)!$ tends to become small very rapidly as m increases. We show numerically that for each fixed f and h—that is, n—there is an optimal finite value of m that minimizes the above error estimate.

The assumption of an infinitely smooth boundary curve was chosen to illustrate the maximum potential of this quadrature rule for our boundary integrals. If this is not the case, then there will be an upper limit on the continuity of $f^{(2m+2)}$ as well. For example, the boundary curve could be defined in terms of piecewise continuous polynomial splines, which would place an upper limit on continuity where the pieces join together. For this situation, the above optimal error estimate would no longer be valid. However, our error estimate analysis can be modified by keeping track of the highest continuity level in $f^{(2m+2)}$.

If the boundary curve has sharp corners, then the trapezoidal rule reduces to second order at best, which is no longer optimal. In this situation, an alternate quadrature procedure—for example, the double exponential quadrature rule—should be implemented, which can take into account the non-smoothness of the boundary curve.

Appendix C
Interior Boundary Value Problem for $D[x,y]$

C.1 Error Analysis for $D[x,y]$

The generalized Fourier series method considered here for an interior problem is constructed by using a dense set of points on an external (auxiliary) curve followed by integration over the domain boundary for each of these points. A finite nicely-spaced subset of these points is selected to generate a finite system of equations leading to an approximate solution. These equations involve the evaluation of integrals arising from the nine-component matrix functions $D[x,y]$ and $P[x,y]$. It is useful for us to analyze the specific integrals generated in each case. However, the complexity of these integrals precludes a complete and unabridged error analysis, so we choose a single component to illustrate the general idea.

Consider $D_{22}[x,y] = D_{22}[x_1,x_2,y_1,y_2]$ for $h = 1/2$, $\lambda = 1$, and $\mu = 1$, defined by

$$
D_{22}[x_1,x_2,y_1,y_2]
$$
$$
= \left(6\pi \left((x_1 - y_1)^2 + (x_2 - y_2)^2 \right)^2 \right)^{-1}
$$
$$
\times \Big(-x_1^4 - 3x_2^4 + 4x_1^3 y_1 - y_1^2(3 + y_1^2)
$$
$$
+ 2x_1 y_1 \left(3 + 2y_1^2 + 4(x_2 - y_2)^2 \right) - x_1^2 \left(3 + 6y_1^2 + 4(x_2 - y_2)^2 \right) + 12x_2^3 y_2
$$
$$
+ (3 - 4y_1^2)y_2^2 - 3y_2^4 + x_2^2(3 - 4y_1^2 - 18y_2^2) + 2x_2 y_2(-3 + 4y_1^2 + 6y_2^2)
$$
$$
+ 12(x_1 - y_1)^2 \left((x_1 - y_1)^2 + (x_2 - y_2)^2 \right)
$$
$$
\times K_0 \left[2 \left((x_1 - y_1)^2 + (x_2 - y_2)^2 \right)^{1/2} \right]
$$
$$
+ 6 \left((x_1 - y_1)^2 + (x_2 - y_2)^2 \right)^{1/2} (x_1 + x_2 - y_1 - y_2)(x_1 - x_2 - y_1 + y_2)
$$
$$
\times K_0 \left[2 \left((x_1 - y_1)^2 + (x_2 - y_2)^2 \right) \right]^{1/2}
$$
$$
- \left((x_1 - y_1)^2 + (x_2 - y_2)^2 \right)^2 \mathrm{Log} \left[(x_1 - y_1)^2 + (x_2 - y_2)^2 \right] \Big),
$$

© Springer Nature Switzerland AG 2020
C. Constanda, D. Doty, *The Generalized Fourier Series Method*, Developments in Mathematics 65, https://doi.org/10.1007/978-3-030-55849-9

which contains both Bessel and logarithmic functions.

Contrary to our case, in a typical boundary integral method, both x and y are on the same curve, which creates singularities that need to be handled by the chosen quadrature method. For this traditional situation, the singularity of $D_{22}[x,y]$ on the boundary is $O[\mathrm{Log}[((x_1-y_1)^2+(x_2-y_2)^2)^{1/2}]]$. We notice that the Bessel functions combine in a way that produces a weak logarithmic singularity. In our method, the point x is on the outer (auxiliary) curve and y is on the inner (domain) boundary. These two curves are separated by a positive minimal distance, so $D_{22}[x,y]$ does not contain singularities. However, it is still possible to get close to a singular condition if the two curves are close to each other at some locations.

C.2 Parametrization of the Integral of $D_{22}[x,y]$

To construct an example and for the sake of simplicity, we take both the inner and outer curves to be concentric circles of radii 1 and 2, respectively, as we have in our problems. We fix the point x on the outer circle to be $(-2,0)$ and parametrize y on the inner circle by

$$\{y_1[t], y_2[t]\} = \{\mathrm{Cos}[t], \mathrm{Sin}[t]\}, \quad 0 \le t < 2\pi.$$

Fixing the point x on the outer circle corresponds to generating individual integral equations. With these choices, the parametrization of $D_{22}[x,y]$ becomes

$$D_{22}\left[-2, 0, y_1[t], y_2[t]\right]$$

$$= -\left(50 + 54\mathrm{Cos}[t]\left(-12K_0\left[2(5+4\mathrm{Cos}[t])^{1/2}\right](2+\mathrm{Cos}[t])^2(5+4\mathrm{Cos}[t])\right)\right.$$

$$+ 6\mathrm{Cos}[2t]\left(-6K_1\left[2(5+4\mathrm{Cos}[t])^{1/2}\right](5+4\mathrm{Cos}[t])^{1/2}\right.$$

$$\times (4 + 4\mathrm{Cos}[t] + \mathrm{Cos}[2t]))$$

$$- 2\mathrm{Cos}[3t] + 33\mathrm{Log}\left[5+4\mathrm{Cos}[t]\right] + 40\mathrm{Cos}[t]\mathrm{Log}\left[5+4\mathrm{Cos}[t]\right]$$

$$\left. + 8\mathrm{Cos}[2t]\mathrm{Log}[5+4\mathrm{Cos}[t]]\right)\left(6\pi(5+4\mathrm{Cos}[t])^2\right)^{-1}.$$

Both the magnitude and smoothness of D_{22} are illustrated in Fig. C.1.

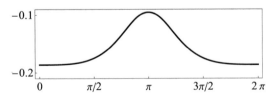

Fig. C.1 Component $D_{22}[-2,0,y_1[t],y_2[t]]$ for $0 \le t \le 2\pi$

The rapid growth of the derivatives of D_{22} can be seen by examining the plot of the sixth derivative of D_{22} in Fig. C.2.

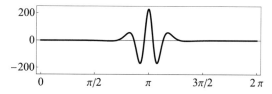

Fig. C.2 Component $d^6 D_{22}[-2,0,y_1[t],y_2[t]]/dt^6$ for $0 \leq t \leq 2\pi$

We remark that the sixth derivative is 3 orders of magnitude larger than the original function. Higher derivatives will grow in magnitude even faster.

For notational convenience, we define

$$D_{22}[t] = D_{22}[-2,0,y_1[t],y_2[t]]$$

and consider the exact value of our test integral

$$\int_0^{2\pi} D_{22}[t]\,dt = -0.998549852118250388\ldots.$$

C.3 The Trapezoidal Quadrature Rule

The above integral, reevaluated by means of the trapezoidal quadrature rule on $n = 50$ equally spaced nodes with 100-digit precision floating-point numbers, yields

$$T_n(\Delta_n(D_{22}[t])) = -0.998549852118250767\ldots.$$

The computational method error of the trapezoidal quadrature rule is

$$T_n(\Delta_n(D_{22}[t])) - \int_0^{2\pi} D_{22}[t]\,dt = -3.78996247\ldots \times 10^{-16}.$$

The error estimate $E_{m,n}[\xi]$ for $m = 2$ and $n = 50$ computed with the sixth derivative of $D_{22}[t]$ is

$$E_{2,50}[\xi] = \frac{2\pi h^6 b_6 D_{22}^{(6)}[\xi]}{6!}.$$

Its graph can be seen in Fig. C.3.

The upper error bound of 2×10^{-7} suggested by $\|E_{2,50}[\xi]\|_\infty$ is far from the actual error, which has magnitude 10^{-16}:

$$\|E_{m,50}[\xi]\|_\infty = (2\pi h^{2m+2}|b_{2m+2}|\|D_{22}^{(2m+2)}\|_\infty)/(2m+2)!.$$

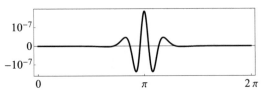

Fig. C.3 $E_{2,50}[\xi]$ for $0 \le \xi \le 2\pi$

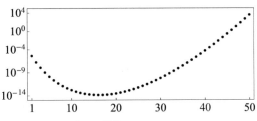

Fig. C.4 $\|E_{m,50}[\xi]\|_\infty$ for $m = 1, 2, 3, \ldots, 50$

Figure C.4 shows the logarithmic plot of the upper error bound as a function of m for $n = 50$.

The upper error bound

$$\operatorname*{Min}_{m \ge 1}\|E_{m,50}[\xi]\|_\infty$$

attains its minimum value of 6.05×10^{-14} at $m = 16$, which is reasonably close to the actual error $-3.789\ldots \times 10^{-16}$. We notice that this bound grows rapidly with m beyond 40.

C.4 Relative Error for the Trapezoidal Quadrature Rule

The relative error for the trapezoidal quadrature rule $T_n(\Delta_n(f))$ as a function of n is defined by

$$E_{\mathrm{Rel}}[n] = \left| \frac{T_n(\Delta_n(f)) - \int_a^b f[t]\,dt}{\int_a^b f[t]\,dt} \right|.$$

For the function $D_{22}[t]$, we calculate $E_{\mathrm{Rel}}[n]$ for $n = 5, 10, 15, 20, \ldots, 200$ and display the result as a logarithmic plot in Fig. C.5.

This graph strongly suggests that the logarithmic relative error is a linear function of n. The relative error can be converted to the form

$$\hat{E}_{\text{Rel}}[n] = O[\text{base}^n].$$

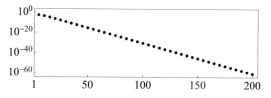

Fig. C.5 $E_{\text{Rel}}[n]$ for $n = 5, 10, 15, 20, \ldots, 200$

Fitting an exponential curve to the data yields

$$\hat{E}_{\text{Rel}}[n] = 0.345773e^{-0.690964n} = 0.3457730 \times 501093^n = 0.345773e^{-4.34146/h}.$$

As seen from Fig. C.6, there is good agreement between the data and the exponential fit.

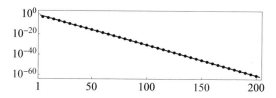

Fig. C.6 $E_{\text{Rel}}[n]$ for $n = 5, 10, \ldots, 200$ and $\hat{E}_{\text{Rel}}[n]$ for $5 \leq n \leq 200$

This numerical example supports our claim that the trapezoidal quadrature rule has exponential precision $O[e^{-4.34146/h}]$ in terms of h, rather than being a fixed-order k method $O[h^k]$. Exponential precision eventually outperforms any fixed-order procedure. For example, consider the fixed orders

$$O[h^k], \quad k = 2, 4, 6, 8,$$

and compare their results with that of the exponential order. Figure C.7 displays the shape of the corresponding curves for the constants in the $O[h^k]$ terms arbitrarily set to 1.

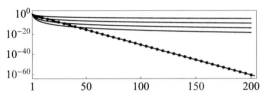

Fig. C.7 $\hat{E}_{\mathrm{Rel}}[n]$ for $5 \leq n \leq 200$ and $O[h^k]$ for $k = 2,4,6,8$

The original non-periodic and non-smooth trapezoidal quadrature rule is at most second order, and Simpson's quadrature rule is at most fourth order. These low-order rules would significantly underperform. For example, if a relative error of 10^{-50} were desired for $D_{22}[t]$, then Simpson's quadrature rule would require approximately 1.03551×10^{13} interpolation points on the domain boundary curve to achieve it. However, the trapezoidal quadrature rule for the periodic and infinitely smooth function $D_{22}[t]$ would require only 165 points. The latter is feasible, whereas the former is not.

It is interesting to see how much additional precision is gained by adding one additional node to the trapezoidal quadrature method. For the above example, we have

$$\text{base} = \frac{\hat{E}_{\mathrm{Rel}}[n+1]}{\hat{E}_{\mathrm{Rel}}[n]} = 0.501093.$$

This indicates that each additional quadrature point reduces the relative error by approximately half for our test integral.

C.5 Effect of the Distance Between Outer (Auxiliary) Curve and Boundary

We recall that the domain boundary

$$\{y_1[t], y_2[t]\} = \{1\mathrm{Cos}[t], 1\mathrm{Sin}[t]\}, \quad 0 \leq t \leq 2\pi,$$

and the outer curve

$$\{x_1[t], x_2[t]\} = \{2\mathrm{Cos}[t], 2\mathrm{Sin}[t]\}, \quad 0 \leq t \leq 2\pi,$$

were chosen to be concentric circles of radii 1 and 2, respectively. The plot of $D_{22}[t] = D_{22}\big[-2, 0, y_1[t], y_2[t]\big]$ with the fixed point $x = (-2,0)$ on the auxiliary circle was shown in Fig. C.1. That graph illustrates the well-behaved smoothness of $D_{22}[t]$ for this particular choice. The test integral was performed over the domain boundary circle $\{y_1[t], y_2[t]\}$ of radius 1, and we see that the result is very well behaved when $r = 2$ for the outer curve. However, if the outer curve has a different radius, then the corresponding quadrature precision will change accordingly.

We define the test integral as a function of the singularity $x = (-r, 0)$ by

$$\int_0^{2\pi} D_{22}\big[-r,0,y_1[t],y_2[t]\big]\,dt,$$

consider six different concentric circles of radii $r = 1.0, 1.2, 1.4, 1.6, 1.8, 2.0$ and six corresponding fixed points $x = (-r,0)$ on those circles, and examine the influence of our choices on $D_{22}\big[-r,0,y_1[t],y_2[t]\big]$ (see Fig. C.8).

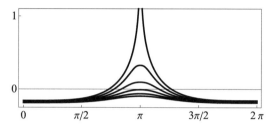

Fig. C.8 $D_{22}\big[-r,0,y_1[t],y_2[t]\big]$ for $r = 1.0, 1.2, 1.4, 1.6, 1.8, 2.0$

We remark that there is a logarithmic singularity when $r = 1$, corresponding to the case where the outer curve touches the domain boundary at $x = (-1,0)$. However, our approximation method precludes this possibility as there is always a minimum positive distance between the two circles. It is also clear that circles closer to the boundary curve will dramatically affect the precision of the trapezoidal quadrature rule because of the expected increase in the magnitude of the higher derivatives of the integrand.

The relative error $E_{\mathrm{Rel}}[n]$ is a function of the number n of equally spaced points t_j used on the domain boundary to construct the trapezoidal quadrature rule $T_n(\Delta_n(f))$, as well as the radius of the outer curve. For $D_{22}[-r,0,y_1[t],y_2[t]]$ and for each outer radii $r = 1.2, 1.4, 1.6, 1.8, 2.0$, we calculated the relative error $E_{\mathrm{Rel}}[n]$ for $n = 5, 10, 15, 20, \ldots, 200$ and displayed the result as a logarithmic plot in Fig. C.9.

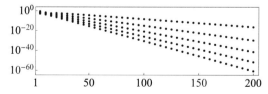

Fig. C.9 $E_{\mathrm{Rel}}[n]$, $n = 5, 10, \ldots, 200$, for $r = 1.2, 1.4, 1.6, 1.8, 2.0$

The data at the top and bottom represent $E_{\mathrm{Rel}}[n]$ for $r = 1.2$ and $E_{\mathrm{Rel}}[n]$ for $r = 2$, respectively. The exponential degree of precision erodes steadily as the outer curve approaches the domain boundary at the point x.

As illustrated earlier, for a radius r and number n of quadrature nodes, the relative error can be modeled by

$$\hat{E}_{\text{Rel}}[r,n] = O[\text{base}[r]^n].$$

For example,

$$\hat{E}_{\text{Rel}}[r = 1.2, n] = 0.208796 \times 0.837599^n, \quad \text{base}[r = 1.2] = 0.837599,$$
$$\hat{E}_{\text{Rel}}[r = 2.0, n] = 0.345773 \times 0.501093^n, \quad \text{base}[r = 2.0] = 0.501093,$$

which indicates a dramatic loss of precision as the outer radius decreases.

We can get a better understanding of base$[r]$ for our specific example. Consider 20 equally spaced radii from $r = 1.1$ to $r = 3.0$, and plot the corresponding values of base$[r]$ vs $1/r$. We expect the trapezoidal quadrature rule convergence to degenerate from exponential to at best quadratic when $r = 1$, because of the loss of continuity of the higher derivatives on the curve when $r = 1$. Hence, we conjecture that exponential convergence completely disappears in this situation, so $\underset{r \to 1^+}{\text{Lim}} \text{base}[r] = 1$. The point $(1, 1)$, displayed with a larger size, is included in Fig. C.10 for reference.

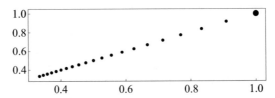

Fig. C.10 base$[r]$ vs $1/r$ for $1.1 \le r \le 3.0$ and base$[1] = 1$

The above plot suggests that base$[r]$ vs $1/r$ for $1 < r$ might be a simple linear function of $1/r$. Since $1 > \text{base}[r] > 0$ for $1 < r < \infty$, a possible model could be

$$\widehat{\text{base}}[r] = O[1/r].$$

Using the above 20 data points produces the simple model

$$\widehat{\text{base}}[r] = 1.003041/r.$$

Superimposing this model on the data indicates a reasonable fit, displayed as $\widehat{\text{base}}[r]$ vs r in Fig. C.11.

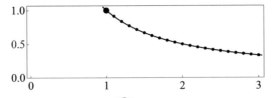

Fig. C.11 Data base$[r]$ and model $\widehat{\text{base}}[r]$ vs r for $1.0 \le r \le 3.0$ and base$[1] = 1$

The simple model $\widehat{\text{base}}[r]$ does an extremely good job of agreeing with the data and with our conjecture that $\text{base}[1] = 1$, because $\widehat{\text{base}}[1] = 1.00304$. It should be noted that this analysis is restricted to just one test integral and should not be interpreted as a guaranteed universal result for other integral equations. However, in what follows we find similar results consistent with these observations.

The above result suggests that it is beneficial for us to choose the outer curve as far away as possible from the domain boundary. However, this generates two conflicting situations. First, for a fixed number of quadrature points on the domain boundary, increasing the outer curve distance from the boundary increases the precision of the quadrature as the higher derivatives of the integrand diminish in magnitude, which improves the degree of precision of the computed system of integral equations. However, increasing the distance between the two curves causes the system of equations to become more ill-conditioned, therefore consuming additional precision in the evaluation. The balance between these two conflicting issues for specific cases is examined later. Finally, the quadrature nodes $\{y_1[t_j], y_2[t_j]\}$ are used to both compute the test integral and to define the domain boundary shape. As the number of such nodes decreases, they become more thinly dispersed and the shape of the domain boundary passing through the nodes $\{y_1[t_j], y_2[t_j]\}$ becomes more ill-defined. This result can be illustrated by observing the scatter that worsens when computing $\hat{E}_{\text{Rel}}[r,n]$ for small values of n.

The above results also suggest that choosing the outer curve too close to the domain boundary can completely undermine computational efficiency. For example, if the outer circle radius is $r = 1.001$, then

$$\hat{E}_{\text{Rel}}[r = 1.001, n] = 28.2636 \times 0.984112^n$$

and $\widehat{\text{base}}[r = 1.001] = 0.984112$. The approximation of the integral of $D_{22}[t]$ by trapezoidal quadrature needs 7397 quadrature nodes to achieve a precision of 10^{-50} when $r = 1.001$. However, only 104 quadrature nodes are required for the same precision if $r = 3$. Since the choice of the outer curve is independent of the boundary integral problem, picking an appropriate one is extremely important.

C.6 Effect of Floating-Point Representational and Computational Precision

The total error in a computed result will be the combined effect of using floating-point numbers as well as the numerical method used to approximate the desired computation. Floating-point representational precision is defined to be the number of digits of precision in the computed result under the assumption that the computational method is perfect. Computational method precision is defined to be the number of digits of precision in the computed result when arithmetic representation is assumed to be perfect. We remark that in most computer floating-point processors, the actual computation may be performed with temporally augmented floating-point representational precision in order to minimize the damaging impact on the computational method precision for each floating-point operation. Typically, the final precision in a computed result is the smaller of the computational method precision and the floating-point representational precision.

Consider the hypothetical representation of a computed floating-point number where the computational method precision is less than the floating-point representational precision; that is,

$$0.d_1d_2d_3\ldots d_cd_{c+1}\ldots d_rd_{r+1}d_{r+2}\ldots \times 10^{n_1n_2\ldots n_m}.$$

The mantissa indicates that the digits $d_1d_2d_3\ldots d_c$ are defined to be free from both computational method and floating-point representational errors. The digits $d_{c+1}\ldots d_rd_{r+1}d_{r+2}\ldots$ are all contaminated by computational method errors. The digits $d_1d_2d_3\ldots d_cd_{c+1}\ldots d_r$ are defined to be free from floating-point representational errors, whereas the digits $d_{r+1}d_{r+2}\ldots$ are not. Consequently, the computational method precision is c digits and the floating-point representation precision is r digits. The significant digits of precision in the final answer will be the smaller of c and r.

The above idea can be illustrated by choosing the trapezoidal quadrature rule approximation of a definite integral as the computational method, and 100-digit floating-point numbers as the floating-point representational precision. It is clear that the two choices are independent of each other, and that either precision can be the smaller of the two.

Consider the computation of our test integral

$$\int_0^{2\pi} D_{22}\big[-2,0,y_1[t],y_2[t]\big]\,dt,$$

which is to be performed over the boundary curve $\{y_1[t],y_2[t]\}$ of radius 1. The computational method is to compute an approximation of this integral using the trapezoidal quadrature rule on 200 equally spaced points. This method alone produces a relative error of 2.99887×10^{-61} in the correct answer.

The quadrature with 200 points was computed repeatedly using floating-point representations ranging from 100 digits down to five digits. The floating-point representational approximation alone produces relative errors in computing the trapezoidal approximation ranging from 8.706×10^{-102} to 1.897×10^{-4}. We notice that the floating-point representational precision in this example is actually slightly higher than the input floating-point precision. This is the result of the specific problem we are solving, and the numerical method used for it. We recall that the magnitude of floating-point representational precision depends in part on the slope of the function being evaluated and, therefore, it is possible in the case of very flat functions to have an actual improvement in their output precision. However, it is much more common that the precision of the computed result is lower (possibly significantly lower) than the input floating-point representational precision.

Figure C.12 contains both the relative computational method precision for $n = 200$ and relative floating-point representational precisions which are displayed versus the number of floating-point digits used.

For this specific example, the relative error of 2.86003×10^{-60} introduced by a floating-point representational precision of 60 digits is about of the same order of magnitude as the relative error of 2.99887×10^{-61} generated by the trapezoidal quadrature rule with 200 points.

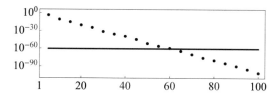

Fig. C.12 Computational and floating-point relative errors

If the objective of the computation is efficiency, then the optimal choice would be to keep both the computational and floating-point errors approximately the same. However, if the objective is to separate the effect of computational method error from floating-point representational error, then the floating-point representational precision must be forced to be higher than the computational method precision. If, for example, in the sequel we intend to isolate and study the computational method error, then we typically choose the floating-point representational precision to be about 150% of the expected computational method precision. This insures the separation of the computational error from the floating-point error.

C.7 Effect of Parametrization

In the previous examples, we took the domain boundary to be a circle of radius 1. Consider the special case where the outer curve is a circle of radius $3/2$ concentric with the boundary. The traditional parametrization of the inner and outer circles are

$$\{y_1[t], y_2[t]\} = \{\mathrm{Cos}[t], \mathrm{Sin}[t]\}, \quad 0 \le t \le 2\pi,$$
$$\{x_1[t], x_2[t]\} = \left\{\tfrac{3}{2}\mathrm{Cos}[t], \tfrac{3}{2}\mathrm{Sin}[t]\right\}, \quad 0 \le t \le 2\pi.$$

These curves can be seen in Fig. C.13, which shows the plot of $D_{22}[-3/2, 0, y_1, y_2]$ with $x(-3/2, 0)$ fixed, as the two solid black lines with the domain removed. This plot allows better visualization of the weak logarithmic singularity at $x = (-3/2, 0)$.

The test integral

$$\int_0^{2\pi} D_{22}\left[-3/2, 0, y_1[t], y_2[t]\right] dt$$

is to be evaluated over the domain boundary $\{y_1[t], y_2[t]\}$. It is clear that the precision of the trapezoidal quadrature rule on this curve deteriorates in the vicinity of the singularity at $x = (-3/2, 0)$ on the outer curve. This loss of precision depends directly on the growth in magnitude of the higher derivatives in the integrand. One possible method to moderate this growth is to re-parametrize the integral. There are a number of different ways to generate a new parametrization, of which we select the most straightforward.

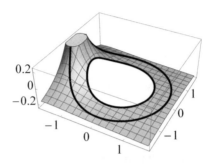

Fig. C.13 Singularity at $x = (-3/2, 0)$ of $D_{22}[-3/2, 0, y_1, y_2]$

We modify the original parametrization

$$0 \le t \le 2\pi$$

by adding a function $g[t]$ such that

$$0 \le t + g[t] \le 2\pi.$$

After a change of variable, the test integral becomes

$$\int_0^{2\pi} D_{22}\big[-3/2, 0, y_1\big[t + g[t]\big], y_2\big[t + g[t]\big]\big](1 + g'[t])\, dt.$$

This change of variable coupled with the continuity of the higher derivatives of D_{22} require the function $g[t]$ to satisfy a number of restrictions:

   all derivatives of the $g[t]$ must be continuous and periodic with period 2π;
   $g[0] = 0, \quad g[2\pi] = 0$;
   $1 + g'[t] \ge 0, \quad 0 \le t \le 2\pi.$

A simple choice for $g[t]$ is the two-parameter function $\beta(\mathrm{Sin}[\alpha\text{-}t]\text{-}\mathrm{Sin}[\alpha])$. We now define the new parametrization $g[\alpha, \beta, t]$ with this $g[t]$ by

$$g[\alpha, \beta, t] = t + \beta(\mathrm{Sin}[\alpha - t] - \mathrm{Sin}[\alpha]), \quad 0 \le t \le 2\pi.$$

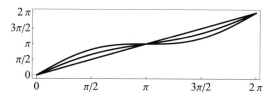

Fig. C.14 Parametrizations $g[\alpha, \beta, t]$ for $\alpha = \pi$, where $\beta = 0, 1/2, 1$

The value of the parameter α for $0 \le \alpha \le 2\pi$ is chosen to be where the outer singularity has the greatest influence on the domain boundary. The parameter β for $0 \le \beta \le 1$ is a tuning parameter used to optimize the precision of the trapezoidal quadrature rule.

The test integral has a singularity at $x = (-3/2, 0)$ on the outer circle, which is in the vicinity of $\alpha = \pi$ on the domain boundary. Figure C.14 illustrates the three parametrizations corresponding to $\beta = 0, 1/2, 1$.

The curve $g[\pi, 1, t]$ has slope 0 at $t = \pi$. This alters the spacing between the quadrature nodes at that location. Consider, for example, 50 equally spaced points t_j in the original parametrization $g[\pi, 0, t] = t$, and plot their new locations in the modified parametrization $g[\pi, 1, t] = t + 1(\text{Sin}[\pi - t] - \text{Sin}[\pi])$, as shown in Fig. C.15.

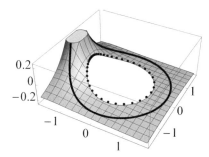

Fig. C.15 $g[\pi, 1, t_j]$ for $j = 1, 2, \ldots, 50$ on $D_{22}[-3/2, 0, y_1, y_2]$

We notice that the modified nodes specified by the parameter $\alpha = \pi$ are concentrated in the vicinity of the singularity at $x = (-3/2, 0)$ on the outer curve . This should improve the accuracy of the trapezoidal quadrature rule.

The relative error $E_{\text{Rel}}[n]$ for the trapezoidal quadrature rule $T_n(\Delta_n(f))$ was defined by

$$E_{\text{Rel}}[n] = \left| \frac{T_n(\Delta_n(f)) - \int_a^b f[t]\, dt}{\int_a^b f[t]\, dt} \right|.$$

Performing a change of variable for the test function D_{22} to the new parametrization $g[\pi, 1, t]$ for $\beta = 1$, we arrive at

$$\hat{D}_{22}[t, \beta = 1] = D_{22}\left[-3/2, 0, \hat{y}_1[t], \hat{y}_2[t]\right]\left(\frac{\partial}{\partial t}g[\pi, 1, t]\right),$$

along the re-parametrized domain boundary

$$\{\hat{y}_1[t], \hat{y}_2[t]\} = \left\{\mathrm{Cos}\left[g[\pi, 1, t]\right], \mathrm{Sin}\left[g[\pi, 1, t]\right]\right\}, \quad 0 \le t \le 2\pi.$$

We calculate $E_{\mathrm{Rel}}[n]$ for the test integral

$$\int_0^{2\pi} \hat{D}_{22}[t, \beta = 1] dt,$$

where $n = 10, 20, 30, \ldots, 200$, and display the result as the logarithmic plot shown in Fig. C.16.

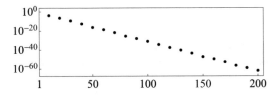

Fig. C.16 $E_{\mathrm{Rel}}[n]$ for parametrization $g[\pi, 1, t]$, $n = 10, 20, 30, \ldots, 200$

Fitting an exponential curve to the data produces

$$\hat{E}_{\mathrm{Rel}}[n] = 0.405547e^{-0.712662n} = 0.405547 \times 0.490337^n = 0.405547e^{-4.47779/h},$$

where, as seen in Fig. C.17, there is good agreement between the data and the exponential fit.

Even for the re-parametrized boundary, the relative error in the trapezoidal quadrature rule exhibits exponential behavior:

$$\hat{E}_{\mathrm{Rel}}[n] = O[\mathrm{base}^n],$$

where the value of base in this special case where $\beta = 1$ is 0.490337. This indicates that for each additional quadrature node added to the boundary, the relative error decreases by a factor of about 49%.

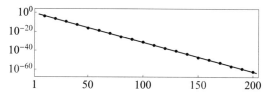

Fig. C.17 $E_{\text{Rel}}[n]$ for $g[\pi, 1, t], n = 10, 20, \ldots, 200$ and $\hat{E}_{\text{Rel}}[n]$ for $10 \leq n \leq 200$

The test integral for arbitrary β, $0 \leq \beta \leq 1$,

$$\int_0^{2\pi} D_{22}\left[-3/2, 0, \hat{y}_1[t], \hat{y}_2[t]\right]\left(\frac{\partial}{\partial t} g[\pi, \beta, t]\right) dt$$

is to be performed over the domain boundary curve

$$\{\hat{y}_1[t], \hat{y}_2[t]\} = \{\text{Cos}\left[g[\pi, \beta, t]\right], \text{Sin}\left[g[\pi, \beta, t]\right]\}, \quad 0 \leq t \leq 2\pi,$$

where

$$g[\pi, \beta, t] = t + \beta(\text{Sin}[\pi - t]).$$

As verified above, the trapezoidal quadrature rule with $\beta = 1$ for this curve with this parametrization is very well behaved. However, different values of the parameter β will produce different accuracies. The primary influence of β is on the growth of the magnitude of the higher derivatives of the integrand $\hat{D}_{22}[t, \beta]$.

We have already stated that the error for the trapezoidal quadrature rule satisfies

$$\left| T_n(\Delta_n(f)) - \int_a^b f[t] \, dt \right| \leq \left(h^{2m+2}(b-a)|b_{2m+2}| \, \|f^{(2m+2)}\|_\infty. \right)/(2m+2)!,$$

where n is the number of equally spaced nodes t_j used on the domain boundary to construct the trapezoidal quadrature rule $T_n(\Delta_n(f))$ for an infinitely smooth periodic function f, and m is any integer such that $m \geq 1$.

We examine the order of magnitude of the higher derivatives $\|f^{(2m+2)}\|_\infty$ for $m = 2$ by studying the sixth derivative of $\hat{D}_{22}[t, \beta]$ for $\beta = 0, 1/2, 1$ (see Figs. C.18, C.19, and C.20).

The case where $\beta = 0$ corresponds to the original unmodified parametrization $g[\pi, 0, t] = t, 0 \leq t \leq 2\pi$. Figure C.18 shows that the order of magnitude of the sixth derivative of $\hat{D}_{22}[t, \beta]$ for $\beta = 0$ in the vicinity of $\alpha = \pi$ is about 10^4.

The case $\beta = 1/2$ corresponds to a parametrization $g[\pi, 1/2, t] = t + \text{Sin}[t]/2$. Figure C.19 shows that the order of magnitude of the sixth derivative of $\hat{D}_{22}[t, \beta]$ for $\beta = 1/2$ in the vicinity of $\alpha = \pi$ is about 50.

The case where $\beta = 1$ corresponds to a parametrization $g[\pi, 1, t] = t + \text{Sin}[t]$. Figure C.20 shows that the order of magnitude of the sixth derivative of $\hat{D}_{22}[t, \beta]$ for $\beta = 1$ in the vicinity of $\alpha = \pi$ is about 400.

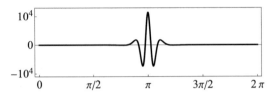

Fig. C.18 Sixth derivative of $\hat{D}_{22}[t, \beta]$ for $\beta = 0$

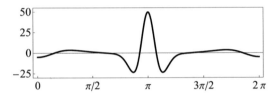

Fig. C.19 Sixth derivative of $\hat{D}_{22}[t, \beta]$ for $\beta = 1/2$

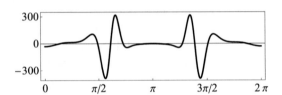

Fig. C.20 Sixth derivative of $\hat{D}_{22}[t, \beta]$ for $\beta = 1$

For the above three cases, the order of magnitude of the sixth derivative is the smallest when the parameter β is 1/2. The influence of β on the higher derivatives actually increases with the order of the derivative. As a consequence, we expect to achieve a significant improvement in precision by using re-parametrization.

We have studied this type of improvement for $\hat{D}_{22}[t, \beta]$ with $\beta = 0, 1/2, 1$ by calculating $E_{\text{Rel}}[n]$ (the relative error in the trapezoidal quadrature rule) for $n = 10, 20, 30, \dots, 200$ and showing the result as the logarithmic plot in Fig. C.21.

The data at the top and bottom represent $E_{\text{Rel}}[n]$ for $\beta = 0$ (that is, the unmodified parametrization), and $E_{\text{Rel}}[n]$ for $\beta = 1/2$, respectively, and the data in the middle corresponds to $E_{\text{Rel}}[n]$ for $\beta = 1$. The exponential degree of precision improves significantly as a function of β. The plot in Fig. C.21 suggests that the optimal value for β is somewhere between 0 and 1. That plot also indicates that for each β and each number n of quadrature nodes, the relative error can be modeled by

$$\hat{E}_{\text{Rel}}[\beta, n] = O[\text{base}[\beta]^n];$$

for example,

$$\hat{E}_{\text{Rel}}[\beta = 0, n] = 0.362971 \times 0.667737^n, \quad \text{base}[\beta = 0] = 0.667737,$$
$$\hat{E}_{\text{Rel}}[\beta = 1/2, n] = 0.460279 \times 0.378923^n, \quad \text{base}[\beta = 1/2] = 0.378923,$$
$$\hat{E}_{\text{Rel}}[\beta = 1, n] = 0.405547 \times 0.490337^n, \quad \text{base}[\beta = 1] = 0.490337.$$

These numbers illustrate a substantial improvement in precision regarded as a function of the parameter β.

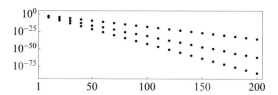

Fig. C.21 $E_{\text{Rel}}[n]$, $n = 10, 20, 30, \ldots, 200$, for $\beta = 0, 1/2, 1$

We can get a better understanding of base$[\beta]$ for our specific test integral. Consider the 20 values $\beta = 0.0, 0.05, 0.1, \ldots, 1.0$ and plot the corresponding values of base$[\beta]$ (see Fig. C.22).

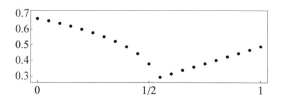

Fig. C.22 base$[\beta]$ for $\beta = 0.0, 0.05, 0.1, \ldots, 1.0$

We notice that the order of $\hat{E}_{\text{Rel}}[\beta, n] = O[\text{base}[\beta]^n]$ is higher for every value of β beyond the non-parametrized value of $\beta = 0$, which indicates that this re-parametrization generates an improvement. We also notice that there are two conflicting trends producing a minimum value of base$[\beta] = 0.294065$ at $\beta = 11/20$. A possible cause for this can be found in Figs. C.18, C.19, and C.20, which suggest that for low values of β, the re-parametrization suppresses the higher derivatives more uniformly over a wider interval of the parameter t, and that this interval narrows as β increases. Also, for higher values of β, the interval of suppression seems to narrow sufficiently to allow a reemergence of the higher order of magnitude of the higher derivatives.

The optimal order of convergence

$$\hat{E}_{\text{Rel}}[\beta, n] = O\big[\text{base}[\beta = 11/20]^n\big] = 0.556417 \times 0.294065^n$$

represents a dramatic level of improvement in the relative error. For example, when $\beta = 11/20$, increasing the number of quadrature nodes by just 1.88 is sufficient to obtain an additional full digit of precision in the quadrature, whereas for $\beta = 0$, an additional 5.70 quadrature nodes would be required to produce the same improvement.

In conclusion, the possible improvement in the order of precision for the relative error in the trapezoidal quadrature rule is nothing short of spectacular. Thus, if the relative error $\hat{E}_{\text{Rel}}[\beta,n]$ in our test integral is limited to 10^{-50} for $\beta = 0$ and $\beta = 11/20$, then $n = 283$ nodes are required for $\beta = 0$ and only $n = 94$ for $\beta = 11/20$ to achieve it. If Simpson's quadrature rule is used for the same test integral and the same order of precision, then $n = 1.91052 \times 10^{50}$ quadrature nodes would be required, which is clearly not feasible.

Appendix D
Exterior Boundary Value Problems for $D^{\mathscr{A}}[x,y]$

D.1 Error Analysis for $D^{\mathscr{A}}[x,y]$

By analogy with the case of the interior problems, the generalized Fourier series method for exterior problems is based on a dense set of points on an inner auxiliary curve and integrals over the domain boundary for each of these points. A finite suitably-spaced subset of these points is selected to generate a finite system of equations that leads to an approximate solution. The algebraic equations derived in this manner are obtained by the evaluation of integrals constructed from the nine component matrix functions $D^{\mathscr{A}}[x,y]$ and $P^{\mathscr{A}}[x,y]$. Below, we intend to analyze the specific integrals generated by the method in each of the external problems. However, the complexity of these integrals makes it impossible for us to carry out a complete and unabridged error analysis, so we confine ourselves to the study of a single component in order to illustrate the general idea.

The component $D^{\mathscr{A}}_{22}[x,y] = D^{\mathscr{A}}_{22}[x_1,x_2,y_1,y_2]$ for $h = 1/2$, $\lambda = 1$, and $\mu = 1$ is

$$
\begin{aligned}
&D^{\mathscr{A}}_{22}[x_1,x_2,y_1,y_2]\\
&= \frac{1}{6\pi}\Bigg(3 - \frac{2x_1^2}{x_1^2+x_2^2} + \mathrm{Log}[x_1^2+x_2^2]\\
&\quad + \big((x_1-y_1)^2+(x_2-y_2)^2\big)^{-2}\big(-x_1^4 - 3x_2^4 + 4x_1^3y_1 - y_1^2(3+y_1^2)\\
&\quad + 2x_1y_1\big(3+2y_1^2+4(x_2-y_2)^2\big) - x_1^2\big(3+6y_1^2+4(x_2-y_2)^2\big) + 12x_2^3y_2\\
&\quad + (3-4y_1^2)y_2^2 - 3y_2^4 + x_2^2(3-4y_1^2-18y_2^2) + 2x_2y_2(-3+4y_1^2+6y_2^2)\\
&\quad + 12(x_1-y_1)^2\big((x_1-y_1)^2+(x_2-y_2)^2\big)\\
&\qquad\qquad\qquad\qquad \times K_0\Big[2\big((x_1-y_1)^2+(x_2-y_2)^2\big)^{1/2}\Big]\\
&\quad + 6\big((x_1-y_1)^2+(x_2-y_2)^2\big)^{1/2}(x_1+x_2-y_1-y_2)(x_1-x_2-y_1+y_2)\\
&\qquad\qquad\qquad\qquad \times B_1\Big[2\big((x_1-y_1)^2+(x_2-y_2)^2\big)^{1/2}\Big]\\
&\quad - \big((x_1-y_1)^2+(x_2-y_2)^2\big)^2\mathrm{Log}[(x_1-y_1)^2+(x_2-y_2)^2]\big)\Bigg);
\end{aligned}
$$

© Springer Nature Switzerland AG 2020
C. Constanda, D. Doty, *The Generalized Fourier Series Method*, Developments in Mathematics 65, https://doi.org/10.1007/978-3-030-55849-9

as can be seen it contains both Bessel and logarithmic functions. We remark that the Bessel functions combine in such a way as to produce a weak logarithmic singularity. In our procedure, the point x is on the inner curve and the point y is on the domain boundary, with the two separated by a positive minimal distance. Hence, $D^{\mathscr{A}}_{22}[x,y]$ has no singularities, although it is still possible to get a singular condition if the two curves are close together at some locations.

D.2 Parametrization of the Integral of $D^{\mathscr{A}}_{22}[x,y]$

To construct an example and for the sake of simplicity, we take both the inner curve and domain boundary to be concentric circles of radii $1/2$ and 1, respectively. We fix the point $x = (-1/2,0)$ on the inner circle and parametrize the point y on the outer circle by

$$\{y_1[t], y_2[t]\} = \{\text{Cos}[t], \text{Sin}[t]\}, \quad 0 \le t \le 2\pi.$$

An exterior problem should have a clockwise parametrization for the test integral. Switching from counterclockwise to clockwise will only switch the sign of the integral. Fixing x on the inner circle corresponds to generating individual integral equations on the boundary. With these choices, the parametrization of the single component $D^{\mathscr{A}}_{22}[x,y]$ becomes

$$D^{\mathscr{A}}_{22}\left[-1/2, 0, y_1[t], y_2[t]\right]$$

$$= -\frac{1}{6\pi}\left(-1 + \text{Log}[4] + (5 + 4\text{Cos}[t])^2\right)^{-2}$$

$$\times \left(65 + 96\text{Cos}[t] - 12K_0[(5 + 4\text{Cos}[t])^{1/2}](1 + 2\text{Cos}[t])^2(5 + 4\text{Cos}[t])\right.$$

$$+ 36\text{Cos}[2t] - 12K_1[(5 + 4\text{Cos}[t])^{1/2}](5 + 4\text{Cos}[t])^{1/2}$$

$$\times (1 + 4\text{Cos}[t] + 4\text{Cos}[2t])$$

$$- 8\text{Cos}[3t] + 33\text{Log}\left[\tfrac{5}{4} + \text{Cos}[t]\right] + 40\text{Cos}[t]\text{Log}\left[\tfrac{5}{4} + \text{Cos}[t]\right]$$

$$+ 8\text{Cos}[2t]\text{Log}\left[\tfrac{5}{4} + \text{Cos}[t]\right]\right).$$

Both the magnitude and smoothness of $D^{\mathscr{A}}_{22}$ are illustrated in Fig. D.1.

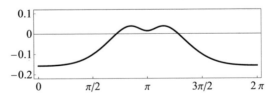

Fig. D.1 Component $D^{\mathscr{A}}_{22}\left[-1/2, 0, y_1[t], y_2[t]\right]$ for $0 \le t \le 2\pi$

The rapid growth of the derivatives of $D_{22}^{\mathscr{A}}$ can be illustrated by examining the plot of the sixth derivative, shown in Fig. D.2.

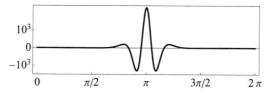

Fig. D.2 Component $d^6 D_{22}^{\mathscr{A}}[-1/2, 0, y_1[t], y_2[t]]/dt^6$ for $0 \leq t \leq 2\pi$

We notice that the sixth derivative is 4 orders of magnitude larger than the original function. Higher derivatives will grow in magnitude even faster.

For notational convenience, we define

$$D_{22}^{\mathscr{A}}[t] = D_{22}^{\mathscr{A}}[-1/2, 0, y_1[t], y_2[t]]$$

and compute the exact value of our test integral:

$$\int_0^{2\pi} D_{22}^{\mathscr{A}}[t]\, dt - -0.476115506107217531\ldots.$$

This integral will be reevaluated by means of the trapezoidal quadrature rule on $n = 50$ equally spaced nodes with 100-digit precision floating-point numbers; that is,

$$T_n\left(\Delta_n(D_{22}^{\mathscr{A}}[t])\right) = -0.476115506107219240\ldots.$$

The error in this case is

$$T_n\left(\Delta_n(D_{22}^{\mathscr{A}}[t])\right) - \int_0^{2\pi} D_{22}^{\mathscr{A}}[t]\, dt = -1.708545233\ldots \times 10^{-15}.$$

D.3 Relative Error for the Trapezoidal Quadrature Rule

The relative error for the trapezoidal quadrature rule $T_n\left(\Delta_n(f)\right)$ as a function of n is defined by

$$E_{\mathrm{Rel}}[n] = \left| \frac{T_n\left(\Delta_n(f)\right) - \int_a^b f[t]\, dt}{\int_a^b f[t]\, dt} \right|.$$

For the function $D^{\mathscr{A}}_{22}[t]$, we calculate $E_{\mathrm{Rel}}[n]$, $n = 5, 10, 15, 20, \ldots, 200$, and display the result as the logarithmic plot shown in Fig. D.3.

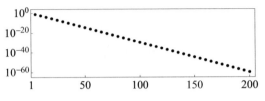

Fig. D.3 $E_{\mathrm{Rel}}[n]$ for $n = 5, 10, 15, 20, \ldots, 200$

This plot strongly suggests that the logarithmic relative error is a linear function of n. Fitting an exponential curve to the data produces

$$\hat{E}_{\mathrm{Rel}}[n] = 3.85281 e^{-0.692644n} = 3.85281 \times 0.500251^n = 3.85281 e^{-4.35201/h}.$$

As Fig. D.4 shows, there is good agreement between the data and the exponential fit.

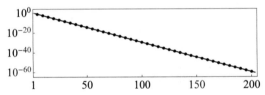

Fig. D.4 $E_{\mathrm{Rel}}[n]$ for $n = 5, 10, \ldots, 200$ and $\hat{E}_{\mathrm{Rel}}[n]$ for $5 \leq n \leq 200$

This numeric example supports the claim that the trapezoidal quadrature rule has exponential precision of $O[e^{-4.35201/h}]$ in terms of h rather than being a fixed-order k method $O[h^k]$. Exponential precision eventually outperforms any fixed-order procedure.

The original non-periodic, non-smooth trapezoidal quadrature rule is second order, whereas Simpson's quadrature rule is at most fourth order. These low-order rules would significantly under-perform. For example, if we wanted to achieve a relative error of 10^{-50} using Simpson's quadrature rule for the integral of $D^{\mathscr{A}}_{22}[t]$, we would require approximately 2.01318×10^{13} interpolation points on the domain boundary, a totally unrealistic proposition. On the other hand, the trapezoidal quadrature rule for the periodic and infinitely smooth function $D^{\mathscr{A}}_{22}[t]$ would require only 168 points, which is very feasible.

We mention that the addition of one extra node to the trapezoidal quadrature method improves its precision considerably. For our example, we get

$$\mathrm{base} = \frac{\hat{E}_{\mathrm{Rel}}[n+1]}{\hat{E}_{\mathrm{Rel}}[n]} = 0.500251,$$

which indicates that each additional quadrature point added to the domain boundary reduces the relative error for our test integral by approximately half.

D.4 Summary

The remaining analysis of the behavior of $D_{22}^{\mathscr{A}}[x,y]$ is essentially unchanged from that performed on $D_{22}[x,y]$ in Appendix C.

Appendix E
Numerical Integration of $P[x,y]$ and $P^{\mathscr{A}}[x,y]$

E.1 Error Analysis for $P[x,y]$ and $P^{\mathscr{A}}[x,y]$

Since $P^{\mathscr{A}}[x,y] = P[x,y]$, it suffices for us to consider only the analysis of an interior problem that makes use of $P[x,y]$. The same reason of complexity of the integrals involved in our method, which we mentioned earlier, makes us choose to investigate a single component of P in order to illustrate the general idea.

The entry $P_{22}[x,y] = P_{22}[x_1,x_2,y_1,y_2]$ of P for $h = 1/2$ and $\lambda = \mu = 1$ is

$$P_{22}[x_1,x_2,y_1,y_2]$$
$$= \left(6\pi\left((x_1 - y_1)^2 + (x_2 - y_2)^2\right)^3\right)^{-1}$$
$$\times \Bigg((x_1 - y_1)\Big(\big(-3 + (x_1 - y_1)^2\big)(x_1 - y_1)^2$$
$$- \big(-9 + (x_2 - y_2)^2\big)(x_2 - y_2)^2\Big)n_1[y_1,y_2]$$
$$+ (x_2 - y_2)\Big(4x_1^4 + 2x_2^4 - 16x_1^3 y_1 - 9y_1^2 + 4y_1^4$$
$$+ 3x_1^2\big(-3 + 8y_1^2 + 2(x_2 - y_2)^2\big)$$
$$- 2x_1 y_1\big(-9 + 8y_1^2 + 6(x_2 - y_2)^2\big)$$
$$- 8x_2^3 y_2 + 3(1 + 2y_1^2)y_2^2 + 2y_2^4$$
$$+ 3x_2^2(1 + 2y_1^2 + 4y_2^2) - 2x_2 y_2(3 + 6y_1^2 + 4y_2^2)\Big)n_2[y_1,y_2]$$
$$+ 6\big((x_1 - y_1)^2 + (x_2 - y_2)^2\big)K_0\Big[2\big((x_1 - y_1)^2 + (x_2 - y_2)^2\big)^{1/2}\Big]$$
$$\times \Big((x_1 - y_1)\big((x_1 - y_1)^2 - 3(x_2 - y_2)^2\big)n_1[y_1,y_2]$$
$$- \big(-3(x_1 - y_1)^2 + (x_2 - y_2)^2\big)(x_2 - y_2)n_2[y_1,y_2]\Big)$$
$$+ 6\big((x_1 - y_1)^2 + (x_2 - y_2)^2\big)^{1/2}K_1\Big[2\big((x_1 - y_1)^2 + (x_2 - y_2)^2\big)^{1/2}\Big]$$
$$\times \Big((x_1 - y_1)\big((1 + (x_1 - y_1)^2)(x_1 - y_1)^2$$
$$- (3 + (x_2 - y_2)^2)(x_2 - y_2)^2\big)n_1[y_1,y_2]$$

© Springer Nature Switzerland AG 2020
C. Constanda, D. Doty, *The Generalized Fourier Series Method*, Developments in Mathematics 65, https://doi.org/10.1007/978-3-030-55849-9

$$+ (x_2 - y_2)\left(-x_2^2 + \left(3 + 2x_2^2 + 2(x_1 - y_1)^2\right)(x_1 - y_1)^2\right.$$
$$+ 2x_2\left(1 - 2(x_1 - y_1)^2\right)y_2$$
$$\left.+ \left(-1 + 2(x_1 - y_1)^2\right)y_2^2\right)n_2[y_1, y_2]\Big)\Big),$$

which contains both Bessel and logarithmic functions, and the outward unit normal $\{n_1[y_1, y_2], n_2[y_1, y_2]\}$ to the boundary curve.

We mention that in a typical boundary integral method, both x and y are on the same curve and therefore give rise to singularities that need to be handled by the chosen quadrature method. For the traditional setup, the singularity of $P_{22}[x,y]$ on the boundary turns out to be of order $O\left[\left((x_1 - y_1)^2 + (x_2 - y_2)^2\right)^{-1/2}\right]$. We note that the combination of the Bessel functions involved creates a strong Cauchy-type singularity. This is quite different than the weak logarithmic singularity of the integral of $D[x,y]$. Since in our approximation method the points x and y are on different curves separated by a minimal positive distance, $P_{22}[x,y]$ contains no singularities. However, it is possible to approach a singular condition if the two curves are close together at some locations. Naturally, we expect the influence of this strong singularity on the accuracy of the trapezoidal quadrature rule to be substantially different than it was for the weak singularity discussed in the previous section.

E.2 Parametrization of the Integral of $P_{22}[x,y]$

To construct an example and for the sake of simplicity, we take both the domain boundary and the outer curve to be concentric circles of radii 1 and 2, respectively. We fix the point $x = (-2,0)$ on the outer circle and parametrize the point y on the inner circle by

$$\{y_1[t], y_2[t]\} = \{\mathrm{Cos}[t], \mathrm{Sin}[t]\}, \quad 0 \le t < 2\pi,$$

with outward unit normal vector to the boundary expressed as

$$\{n_1[y_1[t], y_2[t]], n_2[y_1[t], y_2[t]]\} = \{\mathrm{Cos}[t], \mathrm{Sin}[t]\}.$$

With these choices, the parametrization of $P_{22}[x,y]$ becomes

$$P_{22}[-2, 0, y_1[t], y_2[t]]$$
$$= -\left(12\pi(5 + 4\mathrm{Cos}[t])^3\right)^{-1}$$
$$\times \Big(147 + 190\mathrm{Cos}[t] + 25\mathrm{Cos}[2t] - 30\mathrm{Cos}[3t]$$
$$+ 12K_0[2(5 + 4\mathrm{Cos}[t])^{1/2}](88 + 120\mathrm{Cos}[t] + 33\mathrm{Cos}[2t] + 2\mathrm{Cos}[3t])$$
$$+ 6K_1[2(5 + 4\mathrm{Cos}[t])^{1/2}](5 + 4\mathrm{Cos}[t])^{1/2}$$
$$\times (185 + 262\mathrm{Cos}[t] + 83\mathrm{Cos}[2t] + 10\mathrm{Cos}[3t])$$
$$- 8\mathrm{Cos}[4t]\Big).$$

Both the magnitude and smoothness of P_{22} are illustrated by the plot in Fig. E.1.

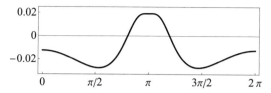

Fig. E.1 Component $P_{22}\big[-2,0,y_1[t],y_2[t]\big]$ for $0 \le t \le 2\pi$

The rapid growth of the derivatives of P_{22} can be seen from the plot of the sixth derivative in Fig. E.2.

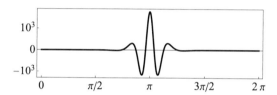

Fig. E.2 Component $d^6P_{22}[-2,0,y_1[t],y_2[t]]/dt^6$ for $0 \le t \le 2\pi$

We notice that the sixth derivative is 5 orders of magnitude larger than the original function. This rate of growth is much higher than that of the sixth derivative of D_{22}. Higher derivatives of P_{22} grow in magnitude even faster.

For notational convenience, we define

$$P_{22}[t] = P_{22}\big[-2,0,y_1[t],y_2[t]\big]$$

and record the exact value of our test integral to be

$$\int_0^{2\pi} P_{22}[t]\,dt = -0.079569659161598\ldots.$$

E.3 The Trapezoidal Quadrature Rule

Our integral will be reevaluated using the trapezoidal quadrature rule on $n = 50$ equally spaced nodes with 100-digit precision floating-point numbers; that is,

$$T_n\big(\Delta_n(P_{22}[t])\big) = -0.0795696591616077\ldots.$$

The computational method error of the trapezoidal quadrature rule is

$$T_n\big(\Delta_n(P_{22}[t])\big) - \int_0^{2\pi} P_{22}[t]\,dt = -0.975104\ldots \times 10^{-14}.$$

The error estimate $E_{m,n}[\xi]$ for $m = 2$ and $n = 50$ which makes use of the sixth derivative of $P_{22}[t]$ is

$$E_{2,50}[\xi] = \frac{2\pi h^6 b_6 P_{22}^{(6)}[\xi]}{6!}.$$

This is graphed in Fig. E.3.

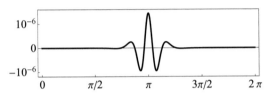

Fig. E.3 $E_{2,50}[\xi]$ for $0 \le \xi \le 2\pi$

The upper error bound of 1.5×10^{-6} suggested by $\|E_{2,50}[\xi]\|_\infty$ is far from the actual error, which has a magnitude of 10^{-14}.

Figure E.4 shows the logarithmic plot of the upper error bound as a function of m for $n = 50$, given by

$$\|E_{m,50}[\xi]\|_\infty = \frac{2\pi h^{2m+2}|b_{2m+2}|\|P_{22}^{(2m+2)}\|_\infty}{(2m+2)!}.$$

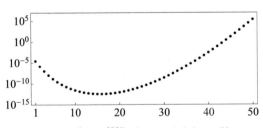

Fig. E.4 $\|E_{m,50}[\xi]\|_\infty$ for $m = 1,2,3,\ldots,50$

The minimum

$$\underset{m\ge 1}{\mathrm{Min}}\|E_{m,50}[\xi]\|_\infty = 3.59301 \times 10^{-13}$$

of the upper error bound occurs at $m = 16$ and is reasonably close to the actual error, which is $-0.975104\ldots \times 10^{-14}$. We notice that the upper error bound grows rapidly with m beyond 40.

E.4 Relative Error for the Trapezoidal Quadrature Rule

The relative error $T_n(\Delta_n(f))$ for the trapezoidal quadrature rule as a function of n is defined by

$$E_{\text{Rel}}[n] = \left| \frac{T_n(\Delta_n(f)) - \int_a^b f[t]\,dt}{\int_a^b f[t]\,dt} \right|.$$

For the function $P_{22}[t]$, we calculate $E_{\text{Rel}}[n]$ with $n = 5, 10, 15, 20, \ldots, 200$. The results are displayed as a logarithmic plot in Fig. E.5.

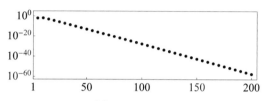

Fig. E.5 $E_{\text{Rel}}[n]$ for $n = 5, 10, 15, 20, \ldots, 200$

The plot strongly suggests that the logarithmic relative error is a linear function of n. This relative error can be converted to the form

$$\hat{E}_{\text{Rel}}[n] = O[\text{base}^n].$$

Fitting an exponential curve to the data yields

$$\hat{E}_{\text{Rel}}[n] = 31.4789e^{-0.67507n} = 31.4789 \times 0.509121^n = 31.4789e^{-4.24159/h}.$$

As seen from Fig. E.6, there is good agreement between the data and the exponential fit.

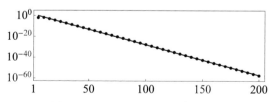

Fig. E.6 $E_{\text{Rel}}[n]$ for $n = 5, 10, \ldots, 200$ and $\hat{E}_{\text{Rel}}[n]$ for $5 \leq n \leq 200$

This numerical example supports the claim that the trapezoidal quadrature rule has exponential precision of $O[e^{-4.24159/h}]$ in terms of h rather than being a method of a fixed-order k with precision $O[h^k]$. Exponential precision eventually outperforms any fixed-order method.

E.5 Effect of the Distance Between Outer (Auxiliary) Curve and Boundary

The component P_{12} generates a plot where the strong singularity at $x = (-2,0)$ is more clearly visible than that of P_{22}; for that reason, we choose to present it here in place of the latter. With our fixed point $x = (-2,0)$, we have

$$
\begin{aligned}
&P_{12}[-2,0,y_1,y_2] \\
&= -\left(6\pi(4+4y_1+y_1^2+y_2^2)^3\right)^{-1} \\
&\quad \times \Big(y_2(20+4y_1-15y_1^2-8y_1^3-y_1^4-3y_2^2+y_2^4)n_1[y_1,y_2] \\
&\qquad + (2+y_1)(20+16y_1^3+2y_1^4+17y_2^2+y_1^2(45+2y_2^2) \\
&\qquad\qquad + y_1(52+8y_2^2))n_2[y_1,y_2] \\
&\quad - 6(4+4y_1+y_1^2+y_2^2)K_0[2(4+4y_1+y_1^2+y_2^2)^{1/2}] \\
&\qquad \times \Big(-y_2\big(-3(2+y_1)^2+y_2^2\big)n_1[y_1,y_2] \\
&\qquad\quad - (2+y_1)(4+4y_1+y_1^2-3y_2^2)n_2[y_1,y_2]\Big) \\
&\quad + 6(4+4y_1+y_1^2+y_2^2)^{1/2}K_1[2(4+4y_1+y_1^2+y_2^2)^{1/2}] \\
&\qquad \times \Big(y_2(-28-44y_1-27y_1^2-8y_1^3-y_1^4+y_2^2+y_2^4)n_1[y_1,y_2] \\
&\qquad\quad - (2+y_1)\big(-4+11y_2^2+2y_2^4+y_1^2(-1+2y_2^2) \\
&\qquad\qquad + y_1(-4+8y_2^2)\big)n_2[y_1,y_2]\Big)\Big),
\end{aligned}
$$

which contains the outward unit normal $\{n_1[y_1,y_2],n_2[y_1,y_2]\}$ to the domain boundary. Unlike $D_{22}[-2,0,y_1,y_2]$, which is defined everywhere, $P_{12}[-2,0,y_1,y_2]$ is restricted to the domain boundary because it involves the normal to the curve. Consequently, we cannot generate an accurate plot of $P_{12}[-2,0,y_1,y_2]$ with a two-dimensional domain. However, since such a plot would be very helpful in visualizing the behavior near the singularity, we have designed an artificial way of extending the definition of the normal vector off the boundary curve. Specifically, each point $\{y_1,y_2\}$ is assumed to lie on the concentric circle passing through it, and we construct our surrogate unit outward normal as the unit outward normal vector to that circle at that point; that is,

$$
\{n_1[y_1,y_2],n_2[y_1,y_2]\} = \left\{\frac{y_1}{(y_1^2+y_2^2)^{1/2}}, \frac{y_2}{(y_1^2+y_2^2)^{1/2}}\right\}.
$$

When $\{y_1,y_2\}$ lies on the boundary circle of radius 1 for parameter t, this reduces to

$$
\{n_1[y_1[t],y_2[t]],n_2[y_1[t],y_2[t]]\} = \{\mathrm{Cos}[t],\mathrm{Sin}[t]\}.
$$

This can be seen in Fig. E.7, which shows the plot of $P_{12}[-2,0,y_1,y_2]$ along with the inner boundary circle of radius 1 and outer circle of radius 2 displayed as solid black lines. The plot, with the interior domain removed, has been rotated counterclockwise for better visualization of the strong singularity at $x = (-2,0)$.

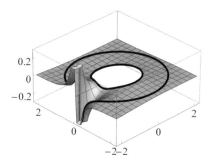

Fig. E.7 The singularity of $P_{12}[-2,0,y_1,y_2]$

Our test integral is now defined as a function of $x = (-r,0)$ by

$$\int_0^{2\pi} P_{12}\big[-r,0,y_1[t],y_2[t]\big]\, dt.$$

The integral, numerically evaluated over the boundary circle $\{y_1[t], y_2[t]\}$, is seen to be very well behaved when the outer circle has radius $r = 2$. However, the selection of a different outer curve will influence the corresponding quadrature precision. We consider six different concentric circles with radii $r = 1.0, 1.2, 1.4, 1.6, 1.8, 2.0$ and six corresponding fixed points $x = (-r,0)$ on those circles, and examine the influence of the various choices on $P_{12}\big[-r,0,y_1[t],y_2[t]\big]$. The graphical results are shown in Fig. E.8.

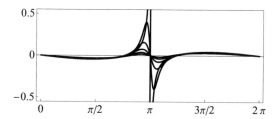

Fig. E.8 $P_{12}\big[-r,0,y_1[t],y_2[t]\big]$ for $r = 1.0, 1.2, 1.4, 1.6, 1.8, 2.0$

$P_{12}\big[-1,0,y_1[t],y_2[t]\big]$ has a strong singularity at $x = (-1,0)$, which completely disappears for $P_{12}\big[-2,0,y_1[t],y_2[t]\big]$ at $x = (-2,0)$. It is clear that circles closer to the boundary significantly affect the precision of the trapezoidal quadrature rule because of the expected increase in the magnitude of the higher derivatives of the integrand.

The component P_{12}, used for visualization of the singularity, is not used to study the performance of the quadrature method because it evaluates to 0, thus causing the relative error $E_{\text{Rel}}[n]$ of our test integral to become undefined. Consequently, our analysis continues with P_{22}.

We choose the test integral

$$\int_{0}^{2\pi} P_{22}\big[-r,0,y_1[t],y_2[t]\big]\,dt.$$

The relative error $E_{\text{Rel}}[n]$ is a function of the number n of equally spaced points t_j used on the domain boundary to construct the trapezoidal quadrature rule $T_n(\Delta_n(f))$, and of the radius r of the outer circle. For $P_{22}\big[-r,0,y_1[t],y_2[t]\big]$ and each of the outer radii $r = 1.2, 1.4, 1.6, 1.8, 2.0$, we calculate $E_{\text{Rel}}[n]$ with $n = 5, 10, 15, 20, \ldots, 200$. The result is displayed as a logarithmic plot in Fig. E.9.

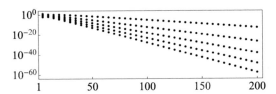

Fig. E.9 $E_{\text{Rel}}[n]$, $n = 5, 10, \ldots, 200$, for $r = 1.2, 1.4, 1.6, 1.8, 2.0$

The data at the top and bottom represent $E_{\text{Rel}}[n]$ for $r = 1.2$ and $E_{\text{Rel}}[n]$ for $r = 2$. The exponential degree of precision erodes steadily as the outer circle approaches the domain boundary. As illustrated previously, for a radius r and number n of quadrature nodes, the relative error can be modeled by

$$\hat{E}_{\text{Rel}}[r,n] = O\big[\text{base}[r]^n\big].$$

For example,

$$\hat{E}_{\text{Rel}}[r = 1.2, n] = 100.223 \times 0.846821^n, \quad \text{base}[r = 1.2] = 0.846821,$$
$$\hat{E}_{\text{Rel}}[r = 2.0, n] = 53.198 \times 0.50717^n, \quad \text{base}[r = 2.0] = 0.50717,$$

which indicates a dramatic loss of precision as the outer radius decreases.

To get a better understanding of $\text{base}[r]$ for our specific example, we consider 20 equally spaced radii from $r = 1.1$ to $r = 3.0$ and plot the corresponding values of $\text{base}[r]$ vs $1/r$. We expect the convergence of the trapezoidal quadrature rule to degenerate from exponential to at best non-exponential quadratic when $r = 1$ because of the loss of continuity of the higher derivatives on the curve. In other words, we expect that $\underset{r \to 1^+}{\text{Lim}}\,\text{base}[r] = 1$, which shows a complete loss of exponential precision. The point $(1,1)$, displayed in a larger size, is included in the plot in Fig. E.10 for reference.

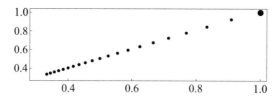

Fig. E.10 base[r] vs $1/r$ for $1.1 \leq r \leq 3.0$ and base[1] = 1

This plot suggests that base[r] vs $1/r$ for $1 < r$ might be a simple linear function of $1/r$. Since $1 > $ base[r] $ > 0$ for $1 < r < \infty$, a possible model could be

$$\widehat{\text{base}}[r] = O[1/r].$$

Using the above 20 data points produces the simple model

$$\widehat{\text{base}}[r] = 1.01249 \, r^{-1}.$$

Superimposing this model on the data indicates a reasonable fit, displayed as $\widehat{\text{base}}[r]$ vs r in Fig. E.11.

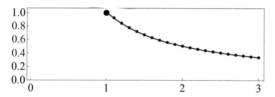

Fig. E.11 Data base[r] and model $\widehat{\text{base}}[r]$ vs r for $1.0 \leq r \leq 3.0$ and base[1] = 1

The simple model $\widehat{\text{base}}[r]$ does an extremely good job of agreeing with the data, as well as with the conjecture that base[1] = 1, given that $\widehat{\text{base}}[1] = 1.01249$. We point out that this analysis is restricted to just one test integral and should not be interpreted as a guaranteed universal result for boundary integral equations. In what follows, we find similar results consistent with these observations.

The above observations suggest that we should choose the outer curve as far away as possible from the domain boundary. However, the comments made about this in Remark 3.9 remain valid here, so we need to make a judicious choice as to how far we should push the outer (auxiliary) circle. Also valid are the comments regarding the ill-defining of the shape of the boundary curve that occurs if the number of the quadrature nodes is too low.

The above results also suggest that choosing the outer curve too close to the domain boundary can completely undermine computational efficiency. For example, taking the outer circle radius to be $r = 1.001$ yields

$$\hat{E}_{\text{Rel}}[r = 1.001, n] = 1.87729 \times 0.999512^n$$

and $\widehat{\text{base}}[r = 1.001] = 0.999512$. Hence, if we wanted to perform trapezoidal quadrature with a precision of 10^{-50}, in this case we would need to take 237190 quadrature nodes to achieve it. As opposed to that, only 110 quadrature nodes would be required for the same precision if $r = 3$ is chosen instead. Since the choice of the outer curve is independent of the boundary integral problem, picking an appropriate outer curve is an extremely important issue.

A comparison between D_{22} (which has a weak logarithmic singularity) and P_{22} (which has a strong singularity) demonstrates the severity of the influence of the latter on quadrature accuracy. As mentioned above, we require 237190 quadrature nodes to achieve a precision of 10^{-50} for P_{22} when $r = 1.001$; the same result for D_{22} is obtained with only 7397 nodes.

E.6 Effect of Parametrization

In the previous examples, we took the boundary to be the circle of radius 1 centered at the origin. Consider the special case where the outer curve has radius $3/2$ and is a circle concentric with the boundary. The traditional parametrization of the outer and inner circles is

$$\{x_1[t], x_2[t]\} = \left\{ \tfrac{3}{2}\text{Cos}[t], \tfrac{3}{2}\text{Sin}[t] \right\}, \quad 0 \le t \le 2\pi,$$
$$\{y_1[t], y_2[t]\} = \{\text{Cos}[t], \text{Sin}[t]\}, \quad 0 \le t \le 2\pi,$$

with corresponding outward unit normal vector

$$\{n_1[y_1[t], y_2[t]], n_2[y_1[t], y_2[t]]\} = \{\text{Cos}[t], \text{Sin}[t]\}.$$

The two curves can be visualized in the plot of $P_{22}[-3/2, 0, y_1, y_2]$ in Fig. E.12 as the two solid black lines with the inner domain removed. The plot allows better visualization of the singularity at $x = (-3/2, 0)$. For P_{22}, it is a strong singularity but in the direction perpendicular to the outer curve and not tangent to it, like the strong singularity in P_{12}. We generated this graph using the artificial procedure described earlier, which includes the temporary modification of the normal vector.

We evaluate the test integral

$$\int\limits_{0}^{2\pi} P_{22}\left[-3/2, 0, y_1[t], y_2[t]\right] dt$$

over the domain boundary $\{y_1[t], y_2[t]\}$.

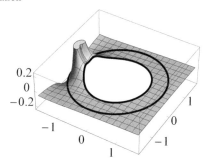

Fig. E.12 Singularity at $x = (-3/2, 0)$ of $P_{22}[-3/2, 0, y_1, y_2]$

It is clear that the precision of the trapezoidal quadrature rule on that curve will deteriorate in the vicinity of the singularity at $x = (-3/2, 0)$ on the outer circle. This loss of precision depends directly on the growth in magnitude of the higher derivatives in the integrand. One possible method to moderate this growth is to re-parametrize the integral, which can be done in a variety of ways. We adopt the most straightforward among them.

Thus, we modify the original parametrization

$$0 \leq t \leq 2\pi$$

by adding a function $g[t]$ such that

$$0 \leq t + g[t] \leq 2\pi.$$

After a change of variable, the test integral becomes

$$\int_0^{2\pi} P_{22}\Big[-3/2, 0, y_1\big[t + g[t]\big], y_2\big[t + g[t]\big]\Big](1 + g'[t])dt.$$

Our change of variable and the continuity of the higher derivatives of P_{22} require the function $g[t]$ to satisfy a few restrictions:

all derivatives of $g[t]$ must be continuous and periodic with period 2π;
$g[0] = 0, \quad g[2\pi] = 0,$
$1 + g'[t] \geq 0, \quad 0 \leq t \leq 2\pi.$

A simple choice for $g[t]$ is the two-parameter function $\beta(\text{Sin}[\alpha - t] - \text{Sin}[\alpha])$. We define the new parametrization by

$$g[\alpha, \beta, t] = t + \beta(\text{Sin}[\alpha - t] - \text{Sin}[\alpha]), \quad 0 \leq t \leq 2\pi.$$

The value of the parameter $\alpha, 0 \leq \alpha \leq 2\pi$, is chosen to be where the outer singularity has the greatest influence on the domain boundary. The parameter $\beta, 0 \leq \beta \leq 1$, is a tuning parameter used to optimize the precision of the trapezoidal quadrature rule.

The test integral has a singularity located at $x = (-3/2,0)$ on the outer circle, which is in the vicinity of $\alpha = \pi$ on the inner boundary curve. Figure E.13 illustrates three parametrizations corresponding to $\beta = 0, 1/2, 1$.

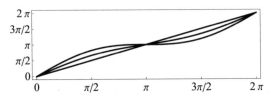

Fig. E.13 Parametrizations $g[\alpha, \beta, t]$ for $\alpha = \pi$ with $\beta = 0, 1/2, 1$

The curve $g[\pi, 1, t]$ has slope 0 at $t = \pi$. This has the effect of altering the spacing between quadrature nodes at that location. We take 50 equally spaced points t_j in the original parametrization $g[\pi, 0, t] = t$ and plot their new locations in the modified parametrization $g[\pi, 1, t] = t + 1(\mathrm{Sin}[\pi - t] - \mathrm{Sin}[\pi])$ (see Fig. E.14).

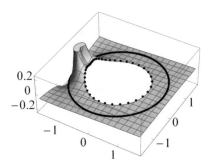

Fig. E.14 $g[\pi, 1, t_j]$ for $j = 1, 2, \ldots, 50$ on $P_{22}[(-3/2,0), y_1, y_2]$

We see that the modified nodes specified by $\alpha = \pi$ are concentrated in the vicinity of the singularity at $x = (-3/2,0)$ on the outer circle. This should improve the accuracy in the trapezoidal quadrature rule.

We go back to the definition of the relative error $E_{\mathrm{Rel}}[n]$ for the trapezoidal quadrature rule $T_n(\Delta_n(f))$ given by the formula

$$E_{\mathrm{Rel}}[n] = \left| \frac{T_n(\Delta_n(f)) - \int_a^b f[t]\,dt}{\int_a^b f[t]\,dt} \right|$$

and perform a change of variable for the test function P_{22} to the new parametrization $g[\pi, 1, t]$ for $\beta = 1$; that is,

$$\hat{P}_{22}[t, \beta = 1] = P_{22}\left[-3/2, 0, \hat{y}_1[t], \hat{y}_2[t]\right]\left(\frac{\partial}{\partial t}g[\pi, 1, t]\right)$$

over the re-parametrized domain boundary

$$\{\hat{y}_1[t], \hat{y}_2[t]\} = \{\text{Cos}\big[g[\pi, 1, t]\big], \text{Sin}\big[g[\pi, 1, t]\big]\}, \quad 0 \le t \le 2\pi,$$

with corresponding outward unit normal vector

$$\{n_1\big[\hat{y}_1[t], \hat{y}_2[t]\big], n_2\big[\hat{y}_1[t], \hat{y}_2[t]\big]\} = \{\text{Cos}\big[t + \text{Sin}[t]\big], \text{Sin}\big[t + \text{Sin}[t]\big]\}.$$

We calculate $E_{\text{Rel}}[n]$ for the test integral

$$\int_0^{2\pi} \hat{P}_{22}[t, \beta = 1]\, dt,$$

where $n = 10, 20, 30, \ldots, 200$. The result is displayed as a logarithmic plot in Fig. E.15.

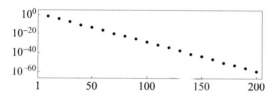

Fig. E.15 $E_{\text{Rel}}[n]$ for the parametrization $g[\pi, 1, t]$ with $n = 10, 20, 30, \ldots, 200$

Fitting an exponential curve to the data produces

$$\hat{E}_{\text{Rel}}[n] = 72.08e^{-0.701852n} = 72.08 \times 0.495667^n = 72.08e^{-4.40986/h}.$$

As seen in Fig. E.16, there is good agreement between the data and the exponential fit.

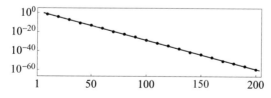

Fig. E.16 $E_{\text{Rel}}[n]$ for $g[\pi, 1, t], n = 10, 20, \ldots, 200, \hat{E}_{\text{Rel}}[n]$ for $10 \le n \le 200$

Even for the re-parametrized boundary, the relative error in the trapezoidal quadrature rule exhibits exponential behavior:

$$\hat{E}_{\text{Rel}}[n] = O[\text{base}^n],$$

where the value of base for the special case of $\beta = 1$ is 0.495667. This indicates that for each additional quadrature node added to the boundary, the relative error decreases by a factor of about 50% when $\beta = 1$.

The test integral for an arbitrary value of β, $0 \leq \beta \leq 1$, is

$$\int_0^{2\pi} P_{22}\big[-3/2,0,\hat{y}_1[t],\hat{y}_2[t]\big]\left(\frac{\partial}{\partial t}g[\pi,\beta,t]\right)dt,$$

which is integrated over the inner boundary curve

$$\{\hat{y}_1[t],\hat{y}_2[t]\} = \big\{\mathrm{Cos}\big[g[\pi,\beta,t]\big],\mathrm{Sin}\big[g[\pi,\beta,t]\big]\big\}, \quad 0 \leq t \leq 2\pi,$$

with corresponding outward unit normal vector

$$\big\{n_1\big[\hat{y}_1[t],\hat{y}_2[t]\big],n_2\big[\hat{y}_1[t],\hat{y}_2[t]\big]\big\} = \big\{\mathrm{Cos}\big[g[\pi,\beta,t]\big],\mathrm{Sin}\big[g[\pi,\beta,t]\big]\big\},$$

where

$$g[\pi,\beta,t] = t + \beta\big(\mathrm{Sin}[\pi - t]\big).$$

It has already been verified that for $\beta = 1$, the trapezoidal quadrature rule for the boundary curve with this parametrization is very well behaved. However, different values of β produce different accuracies. The primary influence of β is to compensate for the growth of the magnitude of the higher derivatives of the integrand $\hat{P}_{22}[t,\beta]$.

We have seen that the error for the trapezoidal quadrature rule satisfies

$$\left|T_n(\Delta_n(f)) - \int_a^b f[t]\,dt\right| \leq \frac{(h^{2m+2}(b-a)|b_{2m+2}|\|f^{(2m+2)}\|_\infty)}{(2m+2)!},$$

where n is the number of equally spaced nodes t_j used on the boundary circle to construct the trapezoidal quadrature rule $T_n(\Delta_n(f))$ for an infinitely smooth periodic function f, and m is any integer such that $m \geq 1$.

The magnitude of the higher derivatives $\|f^{(2m+2)}\|_\infty$ for $m = 2$, examined by studying the sixth derivative of $\hat{P}_{22}[t,\beta]$ for $\beta = 0,1/2,1$, is illustrated in Figs. E.17, E.18, and E.19.

The case where $\beta = 0$ corresponds to the original unmodified parametrization $g[\pi,0,t] = t$, $0 \leq t \leq 2\pi$. We see from Fig. E.17 that the magnitude of the sixth derivative of $\hat{P}_{22}[t,\beta]$ for $\beta = 0$ in the vicinity of $\alpha = \pi$ is about 10^5.

The case $\beta = 1/2$ corresponds to a parametrization of the form

$$g[\pi,1/2,t] = t + \tfrac{1}{2}\mathrm{Sin}[t].$$

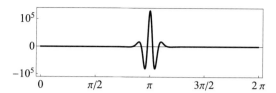

Fig. E.17 Sixth derivative of $\hat{P}_{22}[t,\beta]$ for $\beta = 0$

Figure E.18 shows that the magnitude of the sixth derivative of $\hat{P}_{22}[t,\beta]$ for $\beta = 1/2$ in the vicinity of $\alpha = \pi$ is about 10^2.

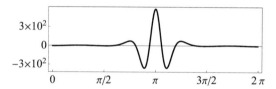

Fig. E.18 Sixth derivative of $\hat{P}_{22}[t,\beta]$ for $\beta = 1/2$

The case $\beta = 1$ corresponds to a parametrization of the form

$$g[\pi,1,t] = t + \mathrm{Sin}[t].$$

Figure E.19 shows that the magnitude of the sixth derivative of $\hat{P}_{22}[t,\beta]$ for $\beta = 1$ in the vicinity of $\alpha = \pi$ is about 10^3.

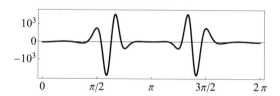

Fig. E.19 Sixth derivative of $\hat{P}_{22}[t,\beta]$ for $\beta = 1$

In the three cases presented above, the magnitude of the sixth derivative is the smallest when $\beta = 1/2$. The influence of the parameter β on the higher derivatives actually increases with the order of the derivative. As a consequence, we expect the use of re-parametrization to bring about a significant improvement in precision.

For $\hat{P}_{22}[t,\beta]$, $\beta = 0, 1/2, 1$, this improvement can be studied by calculating the relative error $E_{\mathrm{Rel}}[n]$ in the trapezoidal quadrature rule with $n = 10, 20, 30, \ldots, 200$. The result is displayed as a logarithmic plot in Fig. E.20.

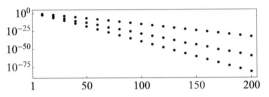

Fig. E.20 $E_{\mathrm{Rel}}[n]$, $n = 10, 20, 30, \ldots, 200$, for $\beta = 0, 1/2, 1$

The data at the top, middle, and bottom represent $E_{\mathrm{Rel}}[n]$ for $\beta = 0$ (i.e. the unmodified parametrization), $E_{\mathrm{Rel}}[n]$ for $\beta = 1$, and $E_{\mathrm{Rel}}[n]$ for $\beta = 1/2$, respectively. The exponential degree of precision improves significantly as a function of β with the optimal value for β suggested by the graph in Fig. E.20 to be somewhere between 0 and 1. That plot also indicates that for each β and each number n of quadrature nodes, the relative error can be modeled by

$$\hat{E}_{\mathrm{Rel}}[\beta, n] = O\big[\mathrm{base}[\beta]^n\big].$$

For example,

$$\hat{E}_{\mathrm{Rel}}[\beta = 0, n] = 57.8997 \times 0.674991^n, \quad \mathrm{base}[\beta = 0] = 0.674991,$$
$$\hat{E}_{\mathrm{Rel}}[\beta = 1/2, n] = 280.6 \times 0.383179^n, \quad \mathrm{base}[\beta = 1/2] = 0.383179,$$
$$\hat{E}_{\mathrm{Rel}}[\beta = 1, n] = 86.6964 \times 0.495013^n, \quad \mathrm{base}[\beta = 1] = 0.495013,$$

which show a dramatic improvement in precision regarded as a function of β.

We can get a better understanding of $\mathrm{base}[\beta]$ for our specific test integral. We consider the 20 values $\beta = 0.0, 0.05, 0.1, \ldots, 1.0$ and examine the plot of the corresponding values of $\mathrm{base}[\beta]$ in Fig. E.21.

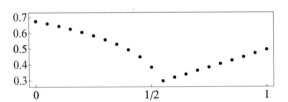

Fig. E.21 $\mathrm{base}[\beta]$ for $\beta = 0.0, 0.05, 0.1, \ldots, 1.0$

The order of $\hat{E}_{\mathrm{Rel}}[\beta, n] = O\big[\mathrm{base}[\beta]^n\big]$ is higher for every value of β beyond the non-parametrized value $\beta = 0$, which suggests that this re-parametrization leads to an improvement. The explanation of the two conflicting trends that produce a minimum value of $\mathrm{base}[\beta] = 0.297305$ at $\beta = 11/20$ and the discussion of the consequences of lower or higher values of β are the same as in Sect. C.7.

The optimal order of convergence

$$\hat{E}_{\text{Rel}}[\beta,n] = O\big[\text{base}[\beta = 11/20]^n\big] = 413.27 \times 0.297305^n$$

represents a significant level of improvement in the relative error. For example, when $\beta = 11/20$, increasing the number of quadrature nodes by just 1.90 nodes is sufficient to obtain an additional full digit of precision in the quadrature, whereas for $\beta = 0$, we would require an additional 5.85 quadrature nodes to achieve the same improvement.

In conclusion, the possible improvement in the order of precision for the relative error in the trapezoidal quadrature rule is quite remarkable. If, say, the relative error $\hat{E}_{\text{Rel}}[\beta,n]$ in our test integral were limited to 10^{-50} for $\beta = 0$ and $\beta = 11/20$, then this outcome would be obtained with $n = 303$ nodes for $\beta = 0$ (the unmodified parametrization), but with only $n = 100$ nodes for $\beta = 11/20$. On the other hand, if we were to use Simpson's quadrature rule for the same test integral and the same order of precision, then we would need $n = 5.05422 \times 10^{50}$ quadrature nodes in the procedure, which is clearly unsustainable.

References

1. Chudinovich, I., Constanda, C.: Variational and Potential Methods in the Theory of Bending of Plates with Transverse Shear Deformation. Chapman & Hall/CRC, Boca Raton, FL (2000)
2. Chudinovich, I., Constanda, C.: Variational and Potential Methods for a Class of Linear Hyperbolic Evolutionary Processes. Springer, London (2004)
3. Chudinovich, I., Constanda, C.: Boundary integral equations in time-dependent bending of thermoelastic plates. J. Math. Anal. Appl. **339**, 1024–1043 (2008)
4. Chudinovich, I., Constanda, C.: The displacement initial-boundary value problem for bending of thermoelastic plates weakened by cracks. J. Math. Anal. Appl. **348**, 286–297 (2008)
5. Chudinovich, I., Constanda, C.: Boundary integral equations in bending of thermoelastic plates with mixed boundary conditions. J. Integral Equations Appl. **20**, 311–336 (2008)
6. Chudinovich, I., Constanda, C.: The traction initial-boundary value problem for bending of thermoelastic plates with cracks. Appl. Anal. **88**, 961–975 (2009)
7. Chudinovich, I., Constanda, C.: Boundary integral equations for thermoelastic plates with cracks. Math. Mech. Solids **15**, 96–113 (2010)
8. Chudinovich, I., Constanda, C.: Boundary integral equations for bending of thermoelastic plates with transmission boundary conditions. Math. Methods Appl. Sci. **33**, 117–124 (2010)
9. Chudinovich, I., Constanda, C.: Transmission problems for thermoelastic plates with transverse shear deformation. Math. Mech. Solids **15**, 491–511 (2010)
10. Chudinovich, I., Constanda, C., Dolberg, O.: On the Laplace transform of a matrix of fundamental solutions for thermoelastic plates. J. Engng. Math. **51**, 199–209 (2005)
11. Chudinovich, I., Constanda, C., Colín Venegas, J.: Solvability of initial-boundary value problems for bending of thermoelastic plates with mixed boundary conditions. J. Math. Anal. Appl. **311**, 357–376 (2005)
12. Chudinovich, I., Constanda, C., Colín Venegas, J.: On the Cauchy problem for thermoelastic plates. Math. Methods Appl. Sci. **29**, 625–636 (2006)
13. Chudinovich, I., Constanda, C., Aguilera-Cortès, L.A.: The direct method in time-dependent bending of thermoelastic plates. Appl. Anal. **86**, 315–329 (2007)
14. Constanda, C.: Direct and Indirect Boundary Integral Equation Methods. Chapman & Hall/CRC, Boca Raton, FL (2000)

© Springer Nature Switzerland AG 2020

C. Constanda, D. Doty, *The Generalized Fourier Series Method*, Developments in Mathematics 65, https://doi.org/10.1007/978-3-030-55849-9

15. Constanda, C.: Mathematical Methods for Elastic Plates. Springer, London (2014)
16. Constanda, C., Doty, D.: Bending of elastic plates: generalized Fourier series method. In: Integral Methods in Science and Engineering: Theoretical Techniques, pp. 71–81 Birkhäuser, New York (2017)
17. Constanda, C., Doty, D.: The Neumann problem for bending of elastic plates. In: Proceedings of the 17th International Conference on Computational and Mathematical Methods in Science and Engineering CMMSE 2017, vol. II, Cádiz, pp. 619–622 (2017)
18. Constanda, C., Doty, D.: Bending of elastic plates with transverse shear deformation: the Neumann problem. Math. Methods Appl. Sci. (2017). https://doi.org/10.1002/mma.4704
19. Constanda, C., Doty, D.: The Robin problem for bending of elastic plates. Math. Methods Appl. Sci. (2018). https://doi.org/10.1002/mma.5286
20. Constanda, C., Doty, D.: Bending of plates with transverse shear deformation: the Robin problem. Comp. Math. Methods, e1015 (2019). https://doi.org/10.1002/cmm4.1015
21. Constanda, C., Doty, D.: Bending of elastic plates: generalized Fourier series method for the Robin problem. In: Integral Methods in Science and Engineering: Analytic Treatment and Numerical Approximations, pp. 97–110. Birkhäuser, New York (2019)
22. Constanda, C., Doty, D.: Analytic and numerical solutions in the theory of elastic plates. Complex Variables Elliptic Equ. (2019). https://doi.org/10.1080/17476933.2019.1636789
23. Constanda, C., Doty, D., Hamill, W.: Boundary Integral Equation Methods and Numerical Solutions: Thin Plates on an Elastic Foundation. Springer, New York (2016)
24. Kirchhoff, G.: Über das Gleichgewicht und die Bewegung einer elastischen Scheibe. J. Reine Angew. Math. **40**, 51–58 (1850)
25. Kreyszig, E.: Introductory Functional Analysis with Applications. Wiley, New York (1978)
26. Kupradze, V.D., Gegelia, T.G., Basheleishvili, M.O., Burchuladze, T.V.: Three-Dimensional Problems of the Mathematical Theory of Elasticity and Thermoelasticity. North-Holland, Amsterdam (1979)
27. Mindlin, R.D.: Influence of rotatory inertia and shear on flexural motions of isotropic elastic plates. J. Appl. Mech. **18**, 31–38 (1951)
28. Reissner, E.: On the theory of bending of elastic plates. J. Math. Phys. **23**, 184–191 (1944)
29. Thomson, G.R., Constanda, C.: Stationary Oscillations of Elastic Plates. A Boundary Integral Equation Analysis. Birkhäuser, Boston (2011)

Index

© Springer Nature Switzerland AG 2020
C. Constanda, D. Doty, *The Generalized Fourier Series Method*, Developments in
Mathematics 65, https://doi.org/10.1007/978-3-030-55849-9

Printed in the United States
by Baker & Taylor Publisher Services